TIM LEWENS is Professor of Philosophy of Science at the University of Cambridge, where he is also a Fellow of Clare College. His past publications include *Organisms and Artifacts: Design in Nature and Elsewhere* (MIT Press, 2004) and *Darwin* (Routledge, 2007). He is also a member of the Nuffield Council on Bioethics and runs the ERC-funded project 'A Science of Human Nature? Philosophical Disputes at the Interface of Natural and Social Sciences'.

The Biological Foundations of Bioethics

The Biological Foundations of Bioethics

Tim Lewens

OXFORD
UNIVERSITY PRESS

OXFORD
UNIVERSITY PRESS

Great Clarendon Street, Oxford, OX2 6DP,
United Kingdom

Oxford University Press is a department of the University of Oxford.
It furthers the University's objective of excellence in research, scholarship,
and education by publishing worldwide. Oxford is a registered trade mark of
Oxford University Press in the UK and in certain other countries

© in this volume Tim Lewens 2015

The moral rights of the author have been asserted

First Edition published in 2015

Impression: 3

All rights reserved. No part of this publication may be reproduced, stored in
a retrieval system, or transmitted, in any form or by any means, without the
prior permission in writing of Oxford University Press, or as expressly permitted
by law, by licence or under terms agreed with the appropriate reprographics
rights organization. Enquiries concerning reproduction outside the scope of the
above should be sent to the Rights Department, Oxford University Press, at the
address above

You must not circulate this work in any other form
and you must impose this same condition on any acquirer

Published in the United States of America by Oxford University Press
198 Madison Avenue, New York, NY 10016, United States of America

British Library Cataloguing in Publication Data
Data available

Library of Congress Control Number: 2014944410

ISBN 978-0-19-871265-7

Printed and bound by
CPI Group (UK) Ltd, Croydon, CR0 4YY

Links to third party websites are provided by Oxford in good faith and
for information only. Oxford disclaims any responsibility for the materials
contained in any third party website referenced in this work.

For Emma

Acknowledgements

The essays collected in this volume were written between 2001 and 2014. Notes in each chapter acknowledge the people who were kind enough to read and comment on particular pieces of work. I also have debts of a more general sort to colleagues and friends who have helped me over longer stretches of time. Within Cambridge I have learned most about the issues examined here from Joanna Burch-Brown, Nick Jardine, Stephen John, Elselijn Kingma, Kathy Liddell, Hallvard Lillehammer, the late Peter Lipton, Neil Manson, Onora O'Neill, Martin Peterson, Martin Richards, Alix Rogers, and Ron Zimmern. I also owe much to the Department of History and Philosophy of Science as a whole (and thereby to John Forrester, Tamara Hug, and David Thompson, who have made the institution run so smoothly), and to Clare College.

Since 2009 I have had the wonderful experience of working as a member of the Nuffield Council on Bioethics. While I have benefited immeasurably from the knowledge of other members of the Council, I am especially grateful to Laura Riley, Hugh Whittall, and Katharine Wright, who, as members of the Council's distinguished secretariat, have deftly helped steer the two working parties I have been involved with, and who have taught me about human bodies and mitochondrial therapies in the process.

Peter Momtchiloff at Oxford University Press has been supportive and efficient throughout the process of putting this book together, and an external reader for OUP offered valuable and generous comments on the newer parts of the manuscript. Work on some of the middle chapters of this book was supported by grants from the Leverhulme Trust and the Isaac Newton Trust. I am also grateful to the European Research Council, whose generous funding under the European Union's Seventh Framework Programme (ERC Grant agreement no. 284123) gave me the time to write all of the more recent chapters (1, 4, 5, 6, and 11), and to compile them into the form presented here. For help with all these tasks, from research to redaction, I have a considerable debt to Beth Hannon, a far more accomplished colleague than anyone could reasonably hope for. Finally, Emma Gilby, to whom these pages are dedicated, has enriched my work and much more besides.

Texts and Permissions

Previously published articles have been changed only in cosmetic ways to correct typographical errors and to ensure consistency in style. I am grateful to the following presses for permissions to re-use my earlier work in this book:

To BMJ Publishing Group for permission to re-use 'Enhancement and Human Nature: The Case of Sandel', first published in *Journal of Medical Ethics* 35 (2009), pp. 354–6.

To Taylor and Francis for permission to re-use 'The Risks of Progress: Precaution and the Case of Human Enhancement', first published in *Journal of Risk Research* 13 (2010), pp. 207–16.

To Springer for permission to re-use 'Human Nature: The Very Idea', first published in *Philosophy and Technology* 25 (2012), pp. 459–74.

To Elsevier for permission to re-use 'From *Bricolage* to Biobricks^{TM}: Synthetic Biology and Rational Design', first published in *Studies in History and Philosophy of Biological and Biomedical Sciences* 44 (2013), pp. 641–8; and 'Development Aid: On Ontogeny and Ethics', first published in *Studies in History and Philosophy of Biological and Biomedical Sciences* 33 (2002), pp. 195–217.

To Cambridge University Press for permission to re-use 'Prospects for Evolutionary Policy', first published in *Philosophy* 78 (2003), pp. 496–514.

To John Wiley and Sons for permission to re-use 'What are Natural Inequalities?', first published in *Philosophical Quarterly* 60 (2010), pp. 264–85.

To Oxford University Press for permission to re-use 'Foot Note', first published in *Analysis* 79 (2010), pp. 468–73.

Contents

1. Introduction: The Biological Foundations of Bioethics	1

Part I. Bettering Nature

2. Enhancement and Human Nature: The Case of Sandel	17
3. The Risks of Progress: Precaution and the Case of Human Enhancement	25
4. Human Nature: The Very Idea	39
5. From *Bricolage* to BioBricks™: Synthetic Biology and Rational Design	60
6. Origins, Parents, and Non-identity	79

Part II. Biology in Ethics and Political Philosophy

7. Development Aid: On Ontogeny and Ethics	95
8. Prospects for Evolutionary Policy	124
9. What Are 'Natural Inequalities'?	144
10. Foot Note	169
11. Health, Naturalism, and Policy	175
References	205
Index	217

1

Introduction

The Biological Foundations of Bioethics

1.1 Bioethics and the Philosophy of Biology

Bioethics and the philosophy of biology are usually considered to be distinct fields, located at some distance from each other on the academic landscape. Philosophers of biology typically have a background in philosophy of science: they are interested in epistemological and metaphysical questions about the nature of species, the interpretation of causal relations in evolutionary theory, the role of model-building in ecological inquiry, and so forth. Little of this touches on issues in medicine, and even less of it has much obvious significance for ethics. Meanwhile, bioethicists come from a motley range of disciplinary backgrounds, but their training is frequently grounded in ethics and political philosophy, perhaps in clinical medicine itself, perhaps in the law. They are squarely focused on normative questions about such things as the relationships between medical professionals and their patients, the just distribution of scarce health resources, or the proper conduct of research projects, and they have little contact with the philosophy of science. There are, of course, notable exceptions to these generalizations. The leading political philosopher of healthcare justice, Norman Daniels, first trained as a philosopher of science (Daniels 2008: 2). The prominent British bioethicist Richard Ashcroft began his career in my own Department of History and Philosophy of Science. John Dupré's justly influential work in the philosophy of biology has always had half an eye on the social and ethical implications of biological research. Several other examples could be given, and there are signs that cross-fertilization is increasing. Among younger scholars it is perhaps less unusual for thinkers to make contributions in both areas: here one might consider the work of Lisa Bortolotti, Matteo Mameli, or Russell Powell. Even so, these remain exceptions for the time being. Indeed, it is not unusual for philosophers of science and bioethicists to be located in entirely different departments within the same university: while philosophers of biology are usually found in faculties of philosophy, or of history and philosophy of science, bioethicists instead have their homes in departments of public health, medicine, law, or in dedicated multidisciplinary centres for practical ethics. The result of all this has been that the two disciplines of bioethics and philosophy of biology have had little contact with each other.

This lack of contact is a matter for regret, as the articles collected together in this volume aim to show. Much work in bioethics, and also work in mainstream ethics and political philosophy, is committed to substantive positions regarding the interpretation of biology. Sometimes these commitments are quite obvious. When, at the end of the last century, Peter Singer outlined a manifesto for a 'Darwinian left', he based his concerns quite explicitly on a belief that left-wing politics needed to be updated in the light of what had been discovered about the ways in which human nature had been shaped by evolution (Singer 1999; see also Chapter 8 of this book). At the other end of the political spectrum, Leon Kass's conservative bioethics frequently makes reference to species' natures (e.g. Kass 1998), and these references are used to justify an ethical opposition to novel uses of reproductive technologies (Chapter 4). Philippa Foot (2001) and Michael Thompson (2008) have both advocated a neo-Aristotelian meta-ethics, according to which our judgements of good and evil make claims that answer to facts about species' natures. Christopher Boorse's seminal account of health judges disease by reference to something Boorse calls the 'general species design' (e.g. Boorse 1977), and Boorse's theory has in turn informed influential work by Norman Daniels on the just allocation of resources that contribute to health (Daniels 1985; 2008). The question for Kass, Thompson, Foot, Boorse, and Daniels is whether modern biology endorses, undermines, or is neutral with regard to these various notions of species' 'natures' and 'the species design' (see Chapters 4, 10, 11).

On other occasions, the commitments of bioethicists and political philosophers to philosophical interpretations of biological reality are less obvious. Norman Daniels is rightly attuned to the difficulty inherent in describing health as a natural state, or a natural good: 'In whatever sense health is a natural good, its distribution is to a large extent socially determined, as is the aggregate level of health in a population' (2008: 13). What is more, Daniels and his collaborators are well aware of the manner in which the apparently 'natural' effects of genes may themselves be contingent on environmental circumstances that are subject to social influence (Buchanan et al. 2000). In spite of all this, they make use of a distinction between natural and social inequalities, a distinction that underlies the work of many other prominent Rawlsians. The notion that some inequalities may be due to nature, and others to society, is one that comes under pressure from more general work within the philosophy of biology that is sceptical of the existence of any good distinction between the natural and the social (see Chapter 9).

For another example of these rather more covert stances regarding biology itself, consider Michael Sandel's work on 'enhancement' technologies (Sandel 2007). Sandel aims to ensure that his opposition to human enhancement does not inadvertently rule out efforts to correct congenital disease. He does so by appealing to a distinction between medical interventions that allow natural capacities to flourish and medical interventions that instead override natural capacities. The former are permissible, the latter are not. But this raises a tricky question: which biological facts make it the case that some of the potentially attainable developmental outcomes for a given individual are to count as instances of the individual flourishing, while others are to count as augmentations that go beyond those limits? Sandel's ethical case against enhancement rests on a position in the philosophy of biology that is both hidden and contestable (see Chapter 2).

In recommending that we assess the biological foundations of bioethics, I am not suggesting simply that bioethicists need to know more basic science if they are to evaluate the desirability of advances in synthetic biology, genetic modification and so forth. Rather, my claim is that ethical discussion often draws on contentious conceptual interpretations of apparently biological facts. Moreover, I do not mean to imply that while philosophers of biology have much to teach people working in ethics, ethicists have nothing to contribute to the philosophy of biology. Valuable influence can run in both directions. To take just one example, albeit one not examined further in this book, consider how one might use research in meta-ethics to inform work on biological function (Lewens 2007a). Simon Blackburn has devoted much of his career to building a sophisticated account of moral utterance that asserts a close linkage between normative judgement and emotional expression, in a manner that aims to evade the objections usually thought to refute earlier versions of emotivism in ethics (e.g. Blackburn 1998). Meanwhile, philosophers of biology have long noted the existence of apparently normative judgements regarding the biological world: in saying that kidneys have the function of filtering blood, many have thought we thereby say something not about what most kidneys are capable of doing, but something about what all kidneys are supposed to do. It is very rare for a philosopher of biology to propose an expressivist analysis of judgements about biological function, perhaps because philosophers of biology have not been attuned to developments in meta-ethics. And yet, Blackburn's form of ethical expressivism offers the possibility of accounting for normativity in a manner that is thoroughly grounded in respectable biological facts, and which also allows that normative judgements can be true or false. It is well worth investigating how far we might apply Blackburn's 'quasi-realism' in the domain of biological function.

Interpretations of biological fact and interpretations of bioethical desirability can exert mutual influences on each other. For a recent example where political debate has affected how embryology is presented, consider the arguments aired in the United Kingdom parliament a few years ago about the acceptability of two experimental techniques that aim at the elimination of disorders of the mitochondrial genome. The mitochondria are small bodies ('organelles') housed outside the nucleus of a cell, within the cytoplasm. Just thirty-seven genes (around 0.1 per cent of the total human genome) are contained within the mitochondria. In spite of their small number, defective mitochondrial genes can lead to a variety of serious conditions, which are often progressive and systemic. Because they are triggered by defects in the mitochondrial genome, the inheritance of these conditions follows a particular pattern: they are passed by mothers (but not by fathers) to their children (both male and female). The first technique for the elimination of these disorders—'pronuclear transfer'—takes a newly formed zygote with defective mitochondria, removes the two pronuclei (i.e. the membrane-bound packages of nuclear chromosomal material deriving from male and female gametes, prior to their fusion to become a single nucleus), and reinserts those pronuclei into a healthy zygote that has had its own two pronuclei removed. 'Maternal spindle transfer' instead involves the removal of the 'spindle' of chromosomes from an unfertilized egg with defective mitochondria. The spindle is then inserted into an (enucleated) egg with healthy mitochondria, and the reconstituted egg is fertilized *in vitro*.[1] The result of both techniques—if they work as intended—is that a woman carrying a mitochondrial disorder can have children who share her (nuclear) genes but who are free from her defective mitochondrial genes, and who are also free from the worrying prospect of passing a mitochondrial disorder to their own children.

It is striking, then, that some UK politicians and scientific commentators have at various times denied that these two types of intervention are properly classed as 'germ-line' therapies. Both interventions have the intended result that a woman who would otherwise have had children and grandchildren affected by mitochondrial disorders will instead have children and grandchildren free from such disorders. In both cases these outcomes are the result of intervening in inherited material contained either within sex cells or within a single-celled zygote. These cells would all typically be thought of as 'germ cells', and the 'germ line' is itself understood simply as a series of such germ cells descending through successive generations. So why deny that these new interventions amount to germ-line therapy? They are, after all, therapeutic and targeted at the germ line.

Lord Walton said in a House of Lords debate on the Human Fertilisation and Embryology Bill:

> People have asked whether this is the same as germ-line gene therapy, a term used for modifying a gene in a nuclear genome at the beginning of development. It is not germ-line therapy, because mitochondrial genomes are not being modified; they are simply being replaced. It is true that, once normal mitochondria are in place, subsequent generations will have normal mitochondria, too, which is hardly a bad thing. (House of Lords Hansard Debates, 3 December 2007)

The then-MP Dr Evan Harris said in a Public Bill Committee Debate on the same bill that 'it is possible to change the DNA in mitochondria

---

[1] For a fuller presentation of the nature of mitochondrial disorders, of both techniques for their elimination and a comprehensive review of the ethical and legal background to these debates, see Nuffield Council on Bioethics (2012). I participated in the drafting of that report as a member of the Council's working party.

without its being considered germ-line gene therapy or germ-line gene engineering, because we restrict that to nuclear DNA' (House of Commons, 3 June 2008).

These are not so much arguments as efforts to stipulate that the term 'germ-line therapy' is to be used either as a label for efforts to change only elements of a genome (as opposed to the whole genome) or for efforts to change only the nuclear genome (as opposed to the mitochondrial genome). It is very hard to see how these stipulations can be justified. Evan Harris might be consoled by the fact that August Weismann eventually came to the view that what he called the 'germ-plasm' was contained entirely within the cell nucleus and not distributed throughout the cell (Winther 2001: 526); however, the very fact that we now recognize significant elements of inherited genetic material to be located in the mitochondria should surely lead us to consider this restriction a mistake on Weismann's part. If we are to argue, with Lord Walton, that germ-line therapies must modify only elements of a genome, then we face the odd conclusion that, if pronuclear and maternal spindle transfer are not germ-line therapies, then replacement of an individual's defective nuclear genome with a healthy genome taken from another individual would not count as a germ-line therapy either, in spite of the lasting effects it would have on that person's children and grandchildren.

Of course, the reason that the likes of Evan Harris and Lord Walton have wanted to deny that the mitochondrial techniques in question are germ-line therapies is presumably because they see no ethical problems associated with them. At the same time, they are aware of the unpleasant moral odour that lingers around interventions in the germ line, as exemplified by the *Universal Declaration on the Human Genome and Human Rights*, which comments that such interventions 'could be contrary to human dignity' (UNESCO 1997, Article 24). But rather than defend these important therapies by denying, implausibly, that they constitute germ-line interventions, it would be more respectable to argue that they demonstrate the ethical defensibility of at least some interventions to the germ line, even if such a defence demands that we take issue with the *Universal Declaration*. It is encouraging, then, that the UK Department of Health's consultation on 'Mitochondrial Donation'—launched in February 2014, and still open as I write these words—is explicit that 'As the aim is that children born as a result of mitochondrial donation, and their offspring, would be free of serious mitochondrial disease, it would...be a form of germ line modification or germ line therapy' (Department of Health 2014: 13). Even here, though, the consultation document's classification of the techniques in question under the heading of 'Mitochondrial Donation' is misleading, for the woman who contributes healthy mitochondria in fact donates every cellular structure bar her nuclear chromosomal material to the woman seeking treatment. This apparently neutral label for the technical facts of the procedures under discussion distracts us from the contributions made to the developing organism by cellular structures that are neither nuclear nor mitochondrial.

These novel procedures illustrate another important benefit of bringing the philosophy of biology into dialogue with bioethics: they make us question the value of terms such as 'germ line' for modern bioethical analysis. As we have seen, the phrase 'germ line' is typically used to name a continuous lineage of 'germ cells'. Commentators have been concerned about interventions in germ cells for a variety of reasons. Some of these reasons have focused on the special moral status often accorded to sex cells and zygotes, by virtue of their originating role in new human life. These reasons also include three far more general grounds for concern. First, commentators have worried that some of these interventions may have long-lasting effects that persist through generations. Second, they have noted that it is in the nature of such interventions that the people whom they will affect cannot consent to them. Third, they have leant on a precautionary fear that the precise effects these interventions will have are both uncertain and potentially harmful to developing children. The fact that all of these concerns also arise when one aims to modify the mitochondrial genome constitutes a further justification for including these therapies under the banner of 'germ-line' interventions; however, at least the more general trio of ethical concerns are also triggered when one intervenes in epigenetic structures within germ cells, and even when one alters long-lasting environmental features and cultural institutions that affect the development of multiple generations. Here, too, we might be concerned that by altering entrenched features of the environment—family structures, early-years educational curricula, nutritional regimes—our interventions can have effects that are uncertain, long-lasting, and outside the bounds of what affected individuals can consent to. It is hard to sustain a raised moral concern for germ-line interventions without tarring town planning with the same brush (see Chapter 7 and Lewens 2004b).

1.2 Overview

The essays collected in this book have all (with the exception of Chapters 6 and 11) been published previously as journal articles. Here they are organized into two broad thematic sections, and within those sections they are printed in the order of their original publication.

The essays in Part I concern the ethics of improving on what nature has given us. Chapters 2 and 3 both concern enhancement, understood here as a label for all efforts to boost human mental and physical capacities beyond what is required for individual health, and ultimately beyond the normal upper range found in our species. Given the breadth of diverse interventions collected under the enhancement umbrella, it is implausible to think that there is any generic ethical case to be made either for or against them. But Michael Sandel has made such a generic case, which focuses on the importance of respecting the 'giftedness' of human nature. In Chapter 2 I argue that Sandel succeeds in diagnosing an important worry we may have about the use of some enhancements by some parents, but his arguments are better understood as opposing 'Procrustean parenting', rather than enhancement in general. Chapter 3 makes a case for thinking that some of the most salient questions about the ethics of enhancement concern the ethics of risk, especially precaution and paternalism. The chapter uses work by John Harris to expose these risk-based considerations, and it aims to counter Harris's enthusiasm for enhancement. More specifically, I argue that a defensible set of precautionary concerns can be isolated, which support scepticism regarding the wisdom of adopting many enhancements in the near and medium-term future. This helps us to justify caution regarding promises of progress in general.

The scepticism of appeals to human nature that underlies Chapter 2 is systematized and elaborated in Chapter 4, where I argue that the only biologically respectable notion of human nature is an extremely permissive one, which names the reliable dispositions of the human species as a whole. This conception has the result that alterations to human nature have been commonplace in the history of our species. I also argue against Aristotelian conceptions of species natures, which are currently fashionable in meta-ethics and applied ethics, on the grounds that they have no basis in biological fact. Moreover, because (as indicated by research in cognitive science) so many of us find this misleading Aristotelian conception highly tempting, I suggest we are better off if we refrain altogether from mentioning human nature in debates over enhancement.

Chapter 5 moves away from enhancement debates, to address a different way of bettering nature. Synthetic biology is often described as a project that applies rational design methods to the organic world. Although generations of humans have influenced organic lineages in many ways, it is nonetheless reasonable to place synthetic biology towards one end of a continuum between purely 'blind' processes of organic modification at one extreme, and wholly rational, design-led processes at the other. I use an example from evolutionary electronics to illustrate some of the constraints imposed by the rational design methodology itself. These constraints remind us of the limitations of the synthetic biology ideal—limitations that are often freely acknowledged by synthetic biology's own practitioners. The synthetic biology methodology, with its emphasis on the need for communication and standardization, is better understood as indicative of an underlying awareness of human limitations, rather than as expressive of an objectionable impulse to mastery over nature.

Chapter 6 wades into murky metaphysical water. The view known as 'origin essentialism' is sometimes thought to entail, or otherwise to offer support for, both gamete essentialism (the view that an organism could not have come from different gametes) and parental essentialism (the view that an organism could not have had different parents). But any plausible version of origin essentialism also needs to respect the intuition that an object's origins could have been slightly different from what they are. Here I suggest that circumstances in which all of a person's original resources are the same except for the gametes that formed her would appear to be circumstances where that same person exists, albeit with slightly different origins. This means that if we are to derive gamete essentialism or parental essentialism from origin essentialism, we need additional contestable premises that explain why the genetic material that goes into a person has special importance among all her other original developmental resources. Our journey into metaphysical obscurity is worthwhile, because this result has significance for non-identity issues when they are raised in bioethics.

Part II of the book addresses a series of more general questions about the relationships between biology, ethics, and politics. Most of its chapters are informed by an interactionist perspective on the processes by which a fertilized egg becomes an adult person. Development is a matter of complex interplay between nutritional regimes, genes, educational regimes, and other diverse developmental resources. In Chapter 7 I suggest there is no ethically salient difference between the contributions made to development by genes and the contributions made by these other resources. Since we think nutrition and schooling should be included in the calculus of distributive justice, we should include at least some genes in this calculus too. What is more, under the right circumstances genetic engineering may become a useful tool for the distribution of developmental resources. This said, attention to the ethical parity of genetic and environmental causation can also help to articulate the legitimate suspicions many groups have of genetic engineering.

Chapter 8 assesses efforts by a minority of biologists, psychologists, and philosophers to show, in various ways, that evolutionary psychology is of relevance to politics and to policy-makers. Two widely accepted arguments suffice (with only a little tweaking) to dismiss such attempts to connect evolution and policy. The first denies the link between adaptation and fixity, the second denies that 'adaptive thinking' is of strong heuristic benefit. Finally, the silence of many evolutionary explanations with respect to developmental mechanisms should also make us suspicious of the relevance of evolutionary psychology to matters of policy.

Chapter 9 addresses a topic of broad relevance to political philosophy. The varying demands of justice are often thought to depend on a distinction between natural and social inequalities, but it is rare to see any explicit account of how this distinction should be drawn. I argue that it cannot be established by a simple causal criterion, nor by the use of the analysis of variance, nor by distinguishing the innate from the acquired. Whether an inequality can be socially controlled provides the most plausible criterion for classing it as social, rather than natural; however, since the question of whether a given inequality is under social control depends in part on how it is described, it is implausible to think that the natural/social distinction is relevant to theories of justice.

Chapter 10 also has broad philosophical relevance, but this time to general work in ethics. According to Philippa Foot (2001), judgements of good or evil are the same in kind as judgements about the goodness or badness of the parts and behaviours of plants or animals. In both cases she takes these judgements to be factual ones, concerning defective functional performance. Foot's case gains prima facie plausibility from the apparently factual nature of judgements regarding the flourishing of plants and animals. Here I argue that the existence of trade-offs between individual survival and reproduction undermines the equation of flourishing with proper organic functioning. These trade-offs present a general challenge for efforts to naturalize ethics by means of naturalized theories of function.

The final long chapter of this book examines the distinction between health and disease, and its relevance to questions in ethics and political philosophy. Any argument to substantiate the ethical salience of the distinction between being healthy and being diseased requires two steps. First, an account must be offered of how to draw the distinction; second, one must show why that distinction marks an important boundary. I argue that naturalistic theories, according to which diseases are understood as biologically malfunctioning traits, fail to make the health/disease distinction ethically salient in itself because of a set of problems involved in trying to derive appropriate goals for medical care from facts about biological goals and the typical effects by which traits contribute to them.

1.3 Three Commitments

Three underlying philosophical commitments unite these twelve essays. First, almost all of them are informed by a general stance that is sceptical both of the ethical significance of claims about what is 'natural', and of the appeals to a substantive notion of human nature (or the nature of other species) on which such claims are often based. This scepticism finds its expression in my rejection of generalized arguments against human enhancement, in my denial of the normative weight sometimes accorded to the health/disease distinction, in my efforts to deflate the prudential significance of claims about what is (and what is not) consonant with our evolved nature, in my fictionalist re-working of the Aristotelian realism of Foot and Thompson, in my analysis of the distinction between natural and social inequality, and doubtless in other places, too.

A second consistent theme that emerges from many of these essays is a general opposition to what is sometimes called 'genetic exceptionalism'. The term 'genetic exceptionalism' has standardly been used to label a view about genetic data: the exceptionalist is one who believes data about our genes to have some kind of special claim to ethical concern, not shared by other forms of personal or medical information (e.g. Richards 2001). I, along with many other writers, am opposed to this kind of genetic exceptionalism; however, when one diagnoses why this narrow form of genetic exceptionalism is so implausible, one is also led to reject a broader exceptionalism that holds not just genetic *data*, but genes more generally, to be loci of special ethical concern. It is a truism of developmental biology that the downstream effects of genes on development are contingent on the environmental background—where this includes both the internal environment of other genes and other biochemicals, and the external environment of natural and cultural resources—that those genes happen to be located in. This means that we can make inferences—albeit inferences that are sometimes shaky—about the presence of genes from their characteristic downstream products. It also means that it is hard even to say what makes a test a genetic test, as opposed to some other form of medical test (Paul 1999). Interventions that allow us to infer the presence of genes may do so by revealing all manner of biomarkers, and they may even proceed in a simple manner by recording family history or easily observable bodily marks. One of the reasons, then, that genetic exceptionalism is so implausible when considered as a view about genetic data is that it is unclear how we are to decide which medical tests are to be subject to special regulation and which are to be accorded a more relaxed form of ethical oversight. The simple interactionist view of development that grounds this scepticism also makes it hard to accord any more general form of special ethical concern to genes, whether that might take the form of opposition to genetic (as opposed to other forms of) enhancement, or whether it might be grounded in the alleged importance of genes (rather than other contributors to development) for identity.

Third, and finally, I have tried throughout these essays to occupy a middle ground in debates over the wisdom of biotechnical progress. We should be suspicious both of the zeal for technological innovation that characterizes many bullish voices within bioethics, and of those forms of technological scepticism that are grounded in indefensible metaphysical images of nature and of the significance of the natural. Utopian visions of a merely imaginable paradise of the neurally enhanced, or of a world where genetic engineering is a ubiquitous tool for the just allocation of basic resources, fail to make any practical case that could legitimate the use of such technologies even in the medium-term future. There remains an enormous amount of work to be done, of a synoptic variety that will bring together research in the social sciences, the natural sciences, and in philosophy, if we want to formulate a more realistic assessment of promises of technical advance.

PART I

Bettering Nature

2

Enhancement and Human Nature
The Case of Sandel

2.1 The Nature of Enhancement

'Human enhancement' is a blanket term that typically refers to a variety of efforts—some still best treated by science fiction, some well-established in today's societies—that are intended to boost our mental and physical capacities, and the capacities of our children, beyond the normal upper range found in our species.[1] Because human enhancement apparently involves altering human nature, it is meant to be the sort of thing that sends shivers down the spine. For 'transhumanists' (e.g. Bostrom 2003), these are frissons of excitement at the thought of a wonderful new world of genetically and pharmaceutically augmented, ultra-intelligent, long-lived super-persons. For conservatives, such as Leon Kass (1998), our shivers are the wise verdict of an instinctive moral repugnance. One of the oddities of this debate is that we have been enhancing human nature for donkeys' years without shivering much at all. Compare, for example, the physical and mental makeup of people in Britain today with the physical and mental makeup of people who lived in Britain 400 years ago. We are taller, we live longer, we have more inclusive ethical codes. These are changes we have wrought on ourselves, and our moral dispositions, our longevity, and our stature are surely elements of human nature. Charles Darwin thought that what he called 'the noblest part of

---

[1] This chapter first appeared in *Journal of Medical Ethics* 35 (2009): 354–6. I am grateful to an audience in Luleå, Sweden, for comments on an early version. Funding was generously provided by the Isaac Newton Trust and the Leverhulme Trust.

our nature'—our sense of sympathy for others—was a product of natural selection (Darwin 1877/2004). But Darwin also thought that this part of our nature had been modified for the better by the deliberate action of humans (Lewens 2007b: ch. 6). Darwin believed that, as our knowledge of the effects of our actions became more detailed, and as we became able to formulate and disseminate public rules of conduct, the scope of sympathy had been altered to encompass not only members of our immediate communities but members of other nations and species. Any evolutionary view that stresses the importance to our species of cultural inheritance (e.g. Richerson and Boyd 2005; Sterelny 2003) will not regard anthropogenic alterations to human nature—including intentional alterations to human nature—as new.

2.2 The Ghost of Eugenics

Before proceeding, we should set one issue aside. Some people might think that, even if enhancement can take many forms, there is something especially disturbing about *genetic* enhancement, for it represents a return to the wrongs of eugenics. But one cannot use parallels with eugenics to justify directing unique moral opprobrium at genetic enhancement (Harris 2007). If one looks at this issue from a historical perspective, one realizes that many eugenicists were concerned with the ways in which a strong genetic inheritance might be squandered in virtue of inadequate educational and nutritional regimes (Kevles 1985; Stepan 1991; Buchanan et al. 2000). Their worries about the health of future generations consequently focused on a variety of interventions—not only genetic ones—that we would now group under the broad banner of public health. From a conceptual standpoint, if one opposes eugenics on the grounds that it embodied state-sponsored efforts to control what sorts of people should exist, then one should also oppose various modern public-health interventions on the same grounds. After all, many governments encourage women who are contemplating conception to take folic acid supplements, thereby reducing the chances of babies being born with spina bifida (Lewens 2004b). One should equally oppose strictly controlled national educational curricula, on the grounds that they also embody an overly restrictive, state-endorsed template for what future people should be like. Let me be clear that this argument is not intended to establish that all of these state-sponsored interventions are legitimate. Rather, it establishes that there is no special link between genetic enhancements and eugenics.

2.3 The Varieties of Enhancement

The ubiquity of enhancement might make us wonder whether there is anything much of a general nature that can be said either for or against it. 'Human nature' is best understood as a name for all the typical features of human populations. Many of its features can be altered in many ways—as many ways as there are of altering developmental processes. 'Enhancement' encompasses experimental infant nutritional regimes, genetic manipulations of the embryo, body building, novel educational practices, the administration of mind-altering drugs, and so forth. In each case, we can examine the goals of these interventions, the mechanisms by which they are achieved, and the likely unintended consequences, and ask whether the ends are worthy, the means appropriate, and the side-effects objectionable. Is it not likely that we will come up with different evaluations in different cases? In line with this, John Harris has made the important point that if we assume that human nature names typical features of human populations, we should conclude that augmenting such capacities as disease resistance or resistance to tooth decay beyond the norm count as augmentations of human nature and are consequently enhancements, in spite of the fact that they are also usually understood to lie within the realm of therapy (Harris 2007). The ubiquity of enhancement constitutes a prima facie case against the idea that 'altering human nature' constitutes any kind of genuine ethical firebreak. And this, in turn, reinforces the thought that there can be no good generic case against enhancement.

2.4 Sandel on Enhancement

Let us recap. There are prima facie reasons to be sceptical of generic cases against enhancement. This does not mean, of course, that we have constructed a case in favour of enhancement. If enhancement is as diverse as I have suggested, we should anticipate that evaluation will need to proceed on a case-by-case basis. But we have not considered all of the detailed cases made against enhancement; perhaps there are legitimate generic worries we should have about enhancement, which the breezy arguments of the preceding sections obscure. In the remainder of this chapter, I consider a recent, sophisticated, and influential case of this sort put forward by Michael Sandel (2007). Sandel also appeals, I will argue, to suspect notions of human nature in laying out his case. He thinks that a diagnosis of what is wrong with enhancement requires us to reconnect with an ethical tradition with which we have lost touch. Specifically, it involves what he calls 'the proper stance of human beings toward the given world' (2007: 9).

Sandel's case rests heavily on this notion of 'the given world', and more specifically on his claim that we should preserve the 'giftedness' of human nature. His devotion to giftedness is inspired in part by theological ethics, but Sandel is careful to note that, as he understands the notion, we need not view life as a gift from God, or from any person. Life is not a gift in the sense of a present that someone has gone to some trouble to bestow on us. Rather, for Sandel, life is a gift in the philosopher's sense of 'the given'. Life is something that a person finds himself with.

Sandel thinks that in seeing life as a gift, we see it as something we ought not to alter, even if we can. Why not? If life were literally a gift from God, then one could understand that there might be good reasons for refusing to fiddle with its makeup. It smacks of ungratefulness. What's more, if God knows best, then tampering with his gift will be counterproductive. But if life is a gift only in the sense that it is something we find ourselves to possess, what is wrong with seeking to alter it, especially if we can alter it for the better? Specifically, why shouldn't parents try to alter the lives of their children for the better? Here is Sandel's answer:

> To appreciate children as gifts is to accept them as they come, not as objects of our design, or products of our will, or instruments of our ambition. Parental love is not contingent on the talents and attributes the child happens to have [... We] do not choose our children. Their qualities are unpredictable, and even the most conscientious parents cannot be held wholly responsible for the kind of child they have. That is why parenthood, more than other human relationships, teaches what the theologian William F. May calls an 'openness to the unbidden'. (2007: 45)

Sandel adds: 'May's resonant phrase describes a quality of character and heart that restrains the impulse to mastery and control and prompts a sense of life as gift' (2007: 46). The 'impulse to mastery' certainly sounds like a bad thing, but it is not clear what is wrong with it. If we are to love our children, it is important that we are disposed to love them however they turn out. Otherwise the chances are we will not love them at all. 'Openness to the unbidden' in this sense is, indeed, a good attitude for parents to have. There is, however, no contradiction in parents being disposed to love their children however they might turn out, while also seeking to influence their children's lives so that they go as well as possible. If 'openness to the unbidden' is to be read as a refusal to intervene in what nature bestows on a child—if this is what a restraint of the impulse to mastery amounts to—then it is no longer clear that it is such an admirable trait. After all, as Harris (2007) has pointed out, it would appear to entail a refusal to allow medical intervention to cure congenital diseases.

Sandel anticipates this problem and clarifies his position further:

> To appreciate children as gifts or blessings is not to be passive in the face of an illness or disease. Healing a sick or injured child does not override her natural capacities but permits them to flourish. Although medical treatment intervenes in nature, it does so for the sake of health, and so does not represent a boundless bid for mastery and dominion. Even strenuous attempts to treat or cure disease do not constitute a Promethean assault on the given. The reason is that medicine is governed, or at least guided, by the norm of restoring and preserving the natural human functions that constitute health. (2007: 46)

Sandel relies here on a slippery distinction between interventions that 'override' natural capacities and those that permit natural capacities to 'flourish'. Consider someone born with the disease phenylketonuria (PKU). The detrimental effects of PKU genes on cognitive development can be greatly eased if the growing child is given a special diet, low in the commonly occurring amino acid phenylalanine, from birth onwards (although see Paul 1998 for some important correctives regarding the ways in which the PKU case has been used by philosophers). Should we say that we have overridden the child's natural capacities by giving her a special diet? Presumably this is instead meant to count as allowing her natural capacities to flourish. But then it becomes hard to see what one might understand by a person's 'natural capacities', beyond those capacities that the person could attain, given the right interventions. Clearly this won't serve to act as a bulwark against enhancement. One might say that genetic alteration is not itself a 'natural' process, hence one cannot describe genetic modification as allowing a person's natural capacities to flourish. But it is equally implausible to think that a specially designed diet, low in phenylalanine, is 'natural'. If 'the given' encompasses all of a person's 'natural capacities', then we do not assault the given either by giving a child a special diet or by giving a child special genes.

2.5 Revisiting the Unbidden

We can understand Sandel's position better—and thereby appreciate what is right about his opposition to enhancement—by returning to his basic ethical case for an 'openness to the unbidden':

> In caring for the health of their children, parents do not cast themselves as designers or convert their children into products of their will or instruments of their ambition [...] Like all distinctions, the line between therapy and enhancement blurs at the edges [...] But this does not obscure the reason the distinction matters: parents bent on enhancing their children are more likely to overreach, to express and entrench attitudes at odds with the norm of unconditional love. (2007: 49)

On the face of things, this doesn't offer much help. It is a defect of character to be disposed to love one's child only if the child turns out a certain way. But a parent who thinks, 'I will only love my child if she has well-above-average intelligence' has the same sort of character fault as a parent who thinks, 'I will only love my child if she has no disabilities or diseases'. The disposition to love conditionally is not the same either as the desire to master nature or as the dismissal of the unbidden. Sandel endorses the efforts of parents who devote considerable time, money, and emotion to seeking cures for their children's diseases. He is not particularly concerned that the actions of such parents are likely to render love conditional on a child being healthy. Why does he think that parents who instead devote time, money, and emotion to enhancing their children's abilities beyond the demands of health will undermine the norm of unconditional love?

We can understand Sandel's reasoning by reflecting on the sorts of case studies he considers. Take the examples of a parent who devotes herself to making her son into a world-leading tennis player and a parent who devotes herself to alleviating the effects of her daughter's cystic fibrosis. One of the most obvious moral differences between the cases originates with the concern that the boy may not care much for tennis. The time and effort he is forced to put into it may deprive him of other activities he would have enjoyed far more, and his relationship with his family may suffer as a result. In brief, the parent may act against the best interests of the child. It is harder—but not impossible—to sketch circumstances under which a parent who acts to relieve a serious disease acts against the interests of her child. Enhancement, more than the curing of disease, tempts parents to shoehorn their children into ill-fitting lives. Enhancement is linked to a genuinely troubling form of 'mastery', not mastery over nature but mastery over the child's emerging character. One can also see why Sandel thinks that this form of 'mastery' is opposed to a valuable form of 'openness to the unbidden'. The good parent will act in concert with the unfolding interests of the growing child, whatever those idiosyncratic interests may turn out to be. The parent who tries to 'master' the child, on the other hand, will seek to promote tennis playing, regardless of what the child turns out to be like. A link now emerges between 'mastery' and conditional love. Consider a parent whose efforts to transform her child into a great tennis player are not tempered by the child's actual inclinations and interests. Such a parent is likely to truly love a child only if the child becomes a great tennis player.

2.6 Procrustean Parenting

There is something right about Sandel's argument. He succeeds in putting his finger on genuine ethical concerns we may have in relation to the efforts of some parents to use some enhancements. But the suggested links between enhancement, conditional love, and insensitivity to the interests of the child are too variable to support a generic case against all forms of enhancement. Not every parent who seeks intensive tennis coaching for her child is insensitive to the child's interests. The child may be an enthusiastic participant in the project, he may thank his parent sincerely for the success he enjoys, it may be that his life would not have gone appreciably better had the parent left him to his own devices. Parents who are attuned to their child's idiosyncrasies could make good use of bespoke enhancements. Moreover, some enhancements seem to have just the same multi-purpose value as the alleviation of serious disease. Long attention span, for example, seems like the kind of trait that is likely to aid a child to achieve her goals and ambitions, whatever those goals might be. (Of course we can imagine very peculiar life goals, the attainment of which would be impeded by long attention span. But we can also imagine very peculiar life goals, the attainment of which would be impeded by being free of serious diseases.) Finally, we can imagine circumstances in which parents put their children through exceptionally demanding therapeutic regimes, when in fact these children—and the society that surrounds them—would be better off finding a way to live with the disease or disability. Efforts to enhance need not go against the interests of the child; efforts to heal need not coincide with the interests of the child. Sandel's case undermines Procrustean parenting, but this is not to say that it rules out all and only cases of enhancement.

Towards the end of his book, Sandel remarks that 'changing our nature to fit the world, rather than the other way around, is actually the deepest form of disempowerment. It distracts us from reflecting critically on the world, and deadens the impulse to social and political improvement' (2007: 97). It is not at all clear what Sandel means by this, partly because it is unclear how we should draw the distinction between the world and our nature. Suppose we begin by noting a feature of the world: children have a variety of different interests, skills, and talents. We also note a feature of our nature: parents are sometimes insufficiently sensitive to their children's idiosyncratic interests. If Sandel's general concerns about the likely effects of enhancement are justified, it can only be because the large majority of parents have this trait. Suppose we all reflect critically on this state of affairs, and the result of that reflection is that we aim to alter this part of our nature. Surely there is nothing wrong with seeking to do this. And if we succeed, a contingent barrier to enhancement is lifted.

3

The Risks of Progress

Precaution and the Case of Human Enhancement

3.1 Introducing Human Enhancement

Progress is supposed to be a good thing.[1] Indeed, anyone who says he is not in favour of progress might well be thought not to understand what 'progress' means. The main goal for this chapter is to use risk-based considerations to justify a form of scepticism about technical progress. I do so by using one specific set of technologies—so-called enhancement technologies—as a case study. Although some of my conclusions are restricted to these technologies, others have quite general significance. They show the legitimacy of precautionary concerns about new technologies, concerns which seek to place high burdens of proof on apparently beneficial technical advances.

I begin by introducing the range of technologies that form the subject matter of this chapter. 'Human enhancement' typically refers to a variety of efforts—some still best treated by science fiction, some well established in today's societies—that are intended to boost our mental and physical capacities, and the capacities of our children, beyond the normal upper range found in our species. Enhancement is sometimes equated with genetic enhancement, but there are plenty of other means by which people might aim—either now or in the future, either successfully or unsuccessfully—to augment their capacities. These include dietary

---

[1] This chapter first appeared in *Journal of Risk Research* 13 (2010): 207–16. An earlier version was presented at the annual Danish/Swedish Risk Meeting in Luleå, Sweden. I am grateful to the audience there, and also to Stephen John, Martin Peterson, Danielle Turner, and especially John Harris for comments on a previous draft. Funding for this work was provided by the Isaac Newton Trust and the Leverhulme Trust.

supplements used to increase long-run resistance to cancer, drugs that enhance athletic performance, medicines that prolong attention span, even meticulously managed 'brain-training' programmes, or intensive educational regimes. Often the same capacity can be boosted in a variety of different ways: memory can be augmented by the practised use of mnemonics, by drugs, and also, in an extended sense, by giving people external data-storage units in the form of mobile phones, computers, or iPods. There is a decent case for thinking that we change human nature by giving humans new external technologies; for those who argue that the skin is a barrier of no special significance to the organism (e.g. Clark and Chalmers 1998; Turner 2000), the fact that augmentation is achieved via external rather than internal modification is not a difference that makes a difference.

3.2 Enhancement and Risk

What, if anything, is wrong with enhancement? The fact that enhancement can take so many forms creates a prima facie case for thinking that there is not much of a general nature that can be said here. Instead, we should decide on the merits of specific potential enhancements by examining the goals of these interventions, the mechanisms by which they are achieved, and their likely intended and unintended consequences; and asking whether the ends are worthy, the means appropriate, and the outcomes and side-effects objectionable. In other words, we should use a variety of considerations relating to risk and evaluate enhancements on a case-by-case basis. But in a rare note of agreement with the conservative Leon Kass (Kass 2003), John Harris writes:

Kass rightly comments that the 'big issues have nothing to do with safety,' for while safety is a big issue it is not a special problem for enhancement technologies or treatments, rather than for, say, nonenhancing therapies. (Harris 2007: 124)

In the remainder of this chapter, I use Harris's recent work on enhancement as a springboard for further reflections about risk and enhancement. Harris is quite right to say that safety, although an important issue for enhancement, is equally important for non-enhancing technologies. Even so, one might invert the significance of his remarks. True enough, there can be intolerable risks associated with proposed therapies, just as there can be intolerable risks associated with proposed enhancements.

But this is compatible with the thought that considerations regarding risk are in fact the most pertinent ones when it comes to assessing the ethics of enhancement technologies. Indeed, I will begin by showing that, whether commentators realize it or not, issues in the philosophy of risk—especially those allied to paternalism and precaution—already underlie debates over enhancement. I will go on to clarify and explore the risk-based issues that are at stake.

3.3 Harris on Enhancement

Harris offers a generic case in favour of enhancement. Its foundation is almost comically simple: 'In terms of human functioning, an enhancement is by definition an improvement on what went before. If it wasn't good for you, it wouldn't be enhancement' (2007: 9). This element of the case is, as Harris puts it, 'trivial', for his definition of 'enhancement' simply prompts debaters to argue about which apparent enhancements are genuine. But Harris's claim parallels a thought one might have about progress in general: just as enhancements are good by definition, so progress is good by definition. If it wasn't an improvement, it wouldn't be progress. And just as one might think only spurious concerns about 'playing God' or 'tampering with human nature' could lead anyone to a general opposition to enhancement, so one might think that only similarly woolly thinking about playing God could underpin general opposition to technical progress.

Harris proceeds to argue that enhancement is not merely permissible, it may be obligatory:

If, as we have suggested, not only are enhancements obviously good for us, but that good can be obtained with safety, then not only should people be entitled to access those goods for themselves and those for whom they care, but they also clearly have moral reasons, perhaps amounting to an obligation, to do so. (Harris 2007: 35)

It is hard to dispute this conclusion, in part because it is framed in a conditional manner: *if* a technology makes our lives, and the lives of our children, better, *then* we have reason to make use of it. Rather than denying this conditional, the sensible opponent of enhancement is likely to confront Harris (or other defenders of enhancement) on different grounds, by containing its practical significance. Although it is rarely described as such, it is at this point that the debate begins to focus on issues relating to the proper handling of risk. For some technologies, sceptics will argue, we have reason to think the antecedent of the conditional is false. And even when we think the antecedent might well be true, a high degree of certainty should be insisted upon before it is accepted as a guide to practical action.

Let me elaborate on these reactions in a little more detail. First, sceptics might feel that, as a matter of fact, most apparent enhancements in fact offer a poor balance of costs and benefits. They are not true enhancements at all, for they will most likely make our lives—and the lives of our children—worse, not better. Second, sceptics might argue that even if it seems plausible that the overall balance sheet of costs and benefits is on the positive side, we should only permit parents to use enhancement technologies if we are entirely sure that their interventions will be in the best interests of the child in question. A demanding burden of proof of this kind will tend to result in highly restricted access to the means of enhancement.

In effect, this amounts to the accusation that defenders of enhancement are guilty of a form of utopianism, which pays too little attention to the realities of enhancement and the uncertainties that surround new technologies (for related thoughts see Ashcroft 2003). Harris encourages this utopian conception of his work at times, pointing out towards the end of his book:

> There is a difference between ethics and public policy. To say that something is ethical and therefore justifiable is not the same as saying it is justified in any particular set of circumstances, nor is it to recommend it nor yet to propose it as a policy for immediate or even for eventual implementation. (Harris 2007: 197)

We might take Harris to be sketching an imagined world not much like our own in which enhancement—because perfectly safe and efficacious—is also obligatory. But this is to undersell Harris's work, for he makes considerable efforts to rebut the sceptic's more practical arguments about risk. Indeed, while Harris explicitly claims, as we have seen, that the 'big issues have nothing to do with safety', the philosophical weight of his case turns on questions about how to handle risk. In rough terms, Harris wishes to argue for three broad sets of claims. First, we should think that many enhancement technologies are genuinely valuable. Second, even if we are unsure of this and suspect that enhancements may be of debatable efficacy or safety, precautionary restrictions on access to them are wrong-headed. Third, the case against precautionary restrictions extends to parental action, which should not be interfered with even when it may cause limited harm to the child. The end result is an attitude that permits enhancements in spite of uncertainty about the overall cost–benefit balance offered by particular technologies.

3.4 From 'Yuck' to 'Wow', and Back Again

Let us begin, then, with the most basic worry that sceptics may have about enhancement, namely, the worry that most apparent enhancements will make our lives worse. Harris's primary ambition for his book is to move his readers' emotional response to enhancement 'from "yuck!" to "wow!"' (2007: 1). It is this element of Harris's case that I will be engaging with in this chapter: my primary concern is to offer considerations supporting those whose attitudes to many enhancements continue, for the moment, to be more 'yuck' than 'wow'. Of course, even if one thought that many enhancements were likely to be mildly dangerous, or of dubious efficacy, one might still claim that the state has no business stopping people from accepting such enhancements, either for themselves or for their children. Let me explain why I do not propose to examine this sort of response in this chapter, even though it needs to be taken seriously and even though it forms a central plank of Harris's case against those who would ban enhancement. Harris frequently points out that we allow parents a broad sphere of free action with regard to the upbringing of their children, in spite of the fact that some activities or interventions encouraged by parents may harm their children in minor ways and even when the overall benefit to children is dubious. Some children hate piano lessons, they may gain no real benefit from them, and they may suffer from being forced to attend. But while we may disapprove of parents who supervise their miserable children's daily practice sessions, few would argue that the state should intervene to prevent this sort of treatment.

Harris illustrates the application of this form of liberty to the enhancement case by asking us to imagine that we could safely equip a child with genes that give it increased patriotism and piety. Some parents might want to equip their children with these genes. Other parents might argue—and Harris seems to endorse this argument—that patriotism and piety are not in the interests of any child to possess. Yet Harris appears to think that prohibiting parents from equipping their children with such genes would be an infringement of their reproductive freedom. This means that even when we suspect that enhancement technologies may go against the interests of children—if, that is, we suspect these technologies may not be genuine enhancements at all—the state should still permit parents to make use of them.

I do not propose to evaluate this argument here, because it does not speak directly to the aim of moving reactions to enhancement from 'yuck' to 'wow'. Harris's anti-paternalistic argument concludes that even if apparent enhancements are mildly harmful, or useless, they should not be banned. That is not to say that specific enhancements are beneficial. Indeed, Harris's own reaction to parents who would instil patriotism and piety in their children appears to be 'yuck'.

In this paper, I therefore leave to one side the complex question of the degree to which the state may be entitled to intervene in parental and personal decisions regarding potentially dangerous enhancements (or treatments, for that matter). Let me be clear, then, that this paper does not show that a ban on enhancement technologies is justified. I focus instead on the more limited question of whether enthusiasm about enhancement is justified. Harris tries to show that it is, by arguing that apparent enhancements are genuine. His most successful arguments in this context are again framed in a conditional manner: *if* we become able to safely boost our mental and physical capacities, *then* we will have good reason to make use of these abilities. Harris is very much in favour of immortality, for example, and he makes a good case against those who argue that if life were greatly extended it would lose its meaning or become intolerably boring. Yet some of Harris's strategies for supporting life extension are self-defeating, at least with respect to his goal of establishing the value of the relevant technologies. For example, he bats away the worries of those who fear overpopulation by claiming that 'cost, risk, and uncertainty will mean that for a very long time the numbers of people availing themselves of such therapies will be a tiny proportion of the world's population' (Harris 2007: 68). In other words, because life-extending measures are likely to be dangerous, expensive, and ambiguous in their benefits, few people will take them up. This hardly shows 'yuck' to be an irrational response, at least not in the short term.

Here Harris's own case in favour of enhancement rests on considerations that also support those sceptics whose refusal to be wowed rests on a worry that many apparent enhancements are likely not to work well, or to be dangerous. These worries have substance. Laboratory experiments have found genetic mutations that dramatically increase the lifespan of fruit flies. But these experiments suggest that the price paid for longer life is reduced fertility (Hughes and Reynolds 2005: 435). Some researchers have suggested the same for the effects of calorie-restricted diets on longevity: their hypothesis is that calorie restriction sends the body into a kind of shock state, in which resources are diverted from the production of gametes into the prolongation of life (Hughes and Reynolds 2005: 434). One day, we might learn how to evade these trade-offs. For the time being, those who put value on the capacity to have children may have good reason for caution regarding some ways of extending life.

In another chapter, Harris praises drugs that enhance cognitive abilities. The drug modafinil—a so-called 'wake-promoting agent', known under the brand name Provigil—has been licensed in the UK for the treatment of narcolepsy since 1997. A 2003 study by Danielle Turner and co-workers concluded that modafinil could also improve the performance of healthy people on some (but not all) of a range of cognitive tasks (Turner et al. 2003). For example, subjects were better able to remember sequences of numbers, which they were asked to repeat first in the order in which they had been presented and then again in reverse order. The group taking a placebo could, on average, recall around 6.5 numbers in the forwards direction and around 5 in the backwards direction. This compared with around 7.7 forwards and 6 backwards for the group taking modafinil. Performance on tasks that involved planning also got better. These improvements are hardly the stuff of *Heroes*, but they are significant improvements all the same. Yet the question of whether modafinil is an overall enhancement remains unresolved. The British Medical Association's November 2007 report *Boosting Your Brainpower* noted that a second study failed to establish any improvement among healthy modafinil users (BMA 2007: 10). The BMA was referring to another 2003 paper, by Delia Randall and colleagues, which concluded (admittedly after investigating half the number of volunteers examined by Turner) that modafinil only affected mood, not cognition (Randall et al. 2003). In fact, this study suggested that the effects of modafinil were negative, appearing to increase both 'psychological anxiety' and 'aggressive mood'.

3.5 Precaution

In the case of modafinil, the scientific evidence supporting the value of this drug as an enhancement remains equivocal. That is not to say, of course, that the drug is in fact harmful, and it certainly does nothing to establish any general scepticism about the value of enhancements. What should we make, then, of the BMA's general concerns with drugs that promote cognitive function? They offer the following caution:

> Although the pharmaceutical products produce interesting and promising results in ideal laboratory conditions, their impact in less controlled situations is still to be investigated. In the meantime, there are risks in attempting to extrapolate from small scale studies. There is some suggestion that improvements in one aspect of an individual's performance may be offset by decreased performance in another aspect. It must also be strongly emphasised here that the side-effects of taking the drugs, particularly over a prolonged period, are unknown and may turn out to be problematic. (BMA 2007: 10)

The so-called 'precautionary principle' has been criticized as being incoherent or vacuous by a number of commentators (e.g. Harris and Holm 2002; Sunstein 2005; Peterson 2006; Manson 2002). These writers are quite right to say that some ways of construing the principle are indefensible, and one might well read the BMA as advocating a vacuous form of precaution here. The BMA is pointing to the possibility of error in moving from laboratory studies to the real world, to the possibility of error in generalizing from small-sample studies, and to the fact that long-term side effects can never be known for sure on the basis of short-term investigation. These things can always be said, even of very well-established therapeutic technologies whose documented benefits are considerable and whose recorded side effects are negligible. Even in these cases one might point out that an inference to their general safety and efficacy is not watertight. But we do not conclude that precautionary restrictions should be placed on well-established therapies merely because it is consistent with our evidence that they might go wrong. We could proceed to argue that any attempt to restrict access to enhancements on precautionary grounds is guilty of double standards: if we are prepared to tolerate fallible inferences from limited samples in the case of established therapies, we should do the same for enhancements.

It is crucial to recognize that not all interpretations of precaution are incoherent (Sandin et al. 2002; Sandin 2007; Stirling 2003; 2005; John 2007; Peterson 2007; Lewens 2008), and there are legitimate precautionary concerns that apply to enhancement technologies. (Note that I do not mean to suggest that Harris would deny any of this.) First, concerns about modafinil are not based merely on the abstract possibility that a finite body of uniformly positive data may be misleading. The Turner study suggests the advantage modafinil gives to healthy users appears to be fairly modest, and when we couple this with indications of negative effects from the Randall study, significant doubt about the overall value of the drug becomes reasonable. Analogous concerns apply to the case of genetic modifications, where the general consensus among biologists is that individual genes are frequently involved in the development of many different traits. The positive effects of genetic modification on the development of one valuable characteristic may therefore be accompanied by negative effects on the development of others. More generally, prospective parents who might consider genetic enhancement need to bear in mind that the technology may offer a poor cost–benefit ratio. This is because valuable traits are typically the result of many different genes acting in concert. Modification of a single gene may only have a very small effect on intelligence or athletic ability. Moreover, the processes required to obtain and manipulate embryos are likely to remain inconvenient, uncomfortable, and expensive for some time. None of these considerations means that all enhancements are, as a matter of fact, dangerous. Even so, they ground a cautious attitude to enhancements in concrete evidence, rather than in abstract concerns about the fallibility of extrapolation from limited data sets.

A second series of legitimate precautionary concerns focuses on the wisdom of taking small steps. In a situation of uncertainty it is wise, wherever possible, to take steps that are either reversible or incremental. If we find we have made a mistake, we can then undo, or failing that limit, the negative results of our actions. This is not, of course, a consideration that applies uniquely to enhancements; it is well established in the case of therapies, too. Consider the Northwick Park trial of the anti-inflammatory drug TGN1412 in March 2006, where the six men who received the active compound suffered from severe adverse events described as 'life-threatening'. It would have been better not to have given the drug to all the volunteers at the same time. Observers would then have been able to halt the trial before all the research subjects were exposed. The same basic thought underlies the current incremental approach to drug regulation, which progresses from tests on animals to tests on small groups of human volunteers, and finally tests on larger groups. As Harris's work also suggests, we should expect the use of apparent enhancements to proceed in a similarly gradual manner, with the possibility of retreat left open as far as possible.

Third, there are reasons to think that there are asymmetries in the evidential burden we should place on the acceptance of enhancements, compared with the acceptance of treatments (here I build on insights from Peterson 2007, and the argument also has resonances with Daniels 2009). Let me introduce this thought by way of analogy. Suppose your watch works poorly. When it runs it is slow and it frequently stops altogether, with the result that you are often confused even about roughly what time it is. Someone offers to improve it, but cautions that the efficacy of the intervention has not been perfectly established, and it may go wrong. Here, you may well accept the intervention. Now suppose your watch keeps time pretty well. It loses a minute every week or so. And someone again offers to improve it so that it loses a minute every year, with the same caveats about the efficacy of the intervention. You are now much less likely to accept the intervention, and before you accept it you should insist on much more evidence backing its efficacy. This is perfectly rational behaviour, because in the first case you have a lot to gain if the intervention goes right and not much to lose if it goes wrong. In the second case the situation is reversed.

We should be wary of applying the watch analogy straightforwardly to the case of human enhancement. I do not claim that the following argument is watertight. Maybe most human beings—even those whose functioning is at the top end of the species distribution—are like broken watches that can easily be made to work better. The moral of the watch story relies on the assumption that as a watch improves, the chances of finding further improvements get lower, while the chances of intervening in ways that make the system function more poorly get higher. This assumption suggests that for a watch that already functions well, further improvements are likely to be of a value that is comparatively small compared with the disvalue that attaches to likely ways in which we might mess the watch up. In contrast, for a watch that functions very badly, the value of likely improvements is more likely to outweigh the disvalue attached to likely foul-ups. Analogous reasoning in the case of human enhancement seems plausible. Someone who is in a state of considerable ill-health will typically have a great deal to gain from therapeutic interventions should they go right. Of course they may also have a great deal to lose from these interventions if they go wrong. But when we look to many enhancements, the gain to someone who is already at a high level of functioning is likely to be rather small, and yet that person still stands to lose considerably if the intervention goes wrong. This will not hold good for all enhancements; it will not hold good, for example, for improvements to human immune systems that allow us to fight off lethal new epidemics. But as a rule of thumb it isn't bad. If your watch, or your body, doesn't work at all, it makes sense for you to take a chance on an untested intervention. If it already works very well indeed, you should demand plenty of evidence before accepting an intervention from someone who says she can make it work even better. And note, importantly, that watches are unlike human organisms in important respects. Watches are made from largely independent parts, so that there will be few systemic knock-on effects if someone tries to tinker with the spring or one of the hands. This restricts concerns about the chances of attempted improvements backfiring. In the case of organisms, the integration of the system means that fiddling with one part can have knock-on consequences for other elements of the system. That does not mean that no enhancements should ever be accepted. But it does make a posture of significant evidential caution a rational one.

These observations are important because they underpin a generic precautionary attitude towards technical improvements. Explaining why this is the case requires a short excursion into theoretical biology. In evolutionary studies it is common to make use of the device of 'fitness landscapes'. In simple cases these take the form of three-dimensional graphs, which allow fitnesses to be plotted for each possible genotype in a population. The horizontal axes represent mutational distances between genotypes, while the vertical axis represents fitness. Small mutational distances correspond to reasonably high-probability genetic alterations, while larger distances correspond to low-probability saltations. Evolution can be depicted as a hill-climbing process on one of these landscapes, whereby some mutations move the population further up adaptive hills, while others move the population down those hills. Fitness landscapes—where the vertical axis represents some measure of functional efficacy, and the horizontal axes measure degrees to which an artefact is modified—can also be used to depict technological change. Some alterations to a device result in functional improvements, or movements up the hill. Others make the device inferior, or move it down the hill. The theoretical biologist Stuart Kauffman has explored the generic properties of fitness landscapes in detail. It is an intuitive feature of these landscapes that as a population moves higher up a hill, the number of directions leading further uphill gets smaller. That, in turn, provides evidence for thinking that the chances of moving further up the hill get smaller. Kauffman puts it this way: 'the higher one is, the harder it is to find a path that continues upward [...] This kind of slowing as fitness increases is a fundamental feature of all adaptive processes [...] This slowing [...] underlies major features of biological and technological evolution' (Kauffman 1996: 168).

Kauffman suggests that there is a dwindling probability of further improvement as adaptation proceeds. We can combine this insight with the way in which the amount of evidence required to confirm the wisdom of some proposed course of action depends on the likely gains and losses that course offers. The result is a justification for setting an increasingly demanding burden of proof the further technology progresses. We justify, in other words, a form of precautionary scepticism regarding technical progress. Note, again, that this does not constitute a watertight argument for this sort of precautionary attitude. The shape of technological fitness landscapes is a contingent matter, and these shapes determine whether it does indeed get progressively harder to find further improvements. If it turns out that the more we come to know, the easier it is to find further improvements, then the argument of the preceding paragraphs does not stand. But a decent prima facie case has, I think, been made.

A fourth and final series of precautionary concerns appeals to broad issues relating to justice and equality. Again, the link with precaution comes via the thought that regulatory decisions need to take full account of all likely impacts of technical change, including impacts relating to justice. This, incidentally, explains one of the rationales for allying participatory approaches to decision-making with precautionary approaches. Thorough risk-based policy should be mindful of how consequences are distributed, and the knock-on effects of those distributions. The benefits of progress are rarely spread equally through society, and many expensive enhancement technologies will doubtless follow the same path. This tendency to promote inequality is not, of course, an intrinsic feature of enhancement, but rather of the distribution of such things as wealth and education, which in turn affect the chances of people acquiring further valuable goods of all kinds. Harris also considers opposition to enhancement based on concerns about equality. He does not deny that enhancement technologies are likely to exacerbate existing inequalities when they first become available, but he argues that it is a good thing if a minority benefits from these technologies, even though many others who could benefit are excluded. Moreover, he argues that by allowing a restricted few to access the benefits of a new technology, we encourage further innovation that will result in price falls and an eventual broadening of availability. Of course, even if Harris is right to argue that it is better for a few to have access to a valuable technology than for none to have access, this does not mean—and Harris would not argue—that the best way to determine access is through ability to pay. We should regret the likely tendency of enhancement to widen wealth-driven inequality. We should also bear in mind important research suggesting that inequality has its own negative effects on health, including the health of those who are better off. It appears, for example, that middle-income groups in unequal societies have lower levels of health than equally affluent (or even poorer) groups in more equal societies (Daniels et al. 2000). Inequality is not just bad for those at the bottom of the pile, and if enhancement promotes inequality it may damage the health of many, including the enhanced. Once again, these are conclusions with generic repercussions for technical advance.

3.6 Conclusion: From Here to Utopia

The case of enhancement is emblematic of more general debates regarding the pitfalls of technical progress. Considerations about risk are important when we consider not merely whether enhancements *could* be ethically sound, but whether enhancements are *likely* to be ethically sound in the near and medium-term future. We can imagine without contradiction a paradise of the enhanced, in which all have the physical and mental capacities needed for a long life of devotion to a broad range of valuable and rewarding pursuits. Harris is right to say that such a utopia is wow-worthy. But for the time being, issues relating to risk—and more specifically issues relating to safety, efficacy, scientific uncertainty, and inequality—justify those who greet the prospect of many enhancements with a muted 'yuck' at best.

4
Human Nature
The Very Idea

4.1 Improving Unicorns?

In bioethical circles there has been plenty of debate in recent years about the wisdom of attempts to alter human nature (e.g. Habermas 2003; Savulescu and Bostrom 2009; Sandel 2007; Harris 2007).[1] According to some philosophers of biology (and biologists), this is rather like asking about the wisdom of altering a unicorn. There's no such thing as a unicorn, and, some have said, there's no such thing as human nature, either. The distinguished philosopher David Hull was consistently 'suspicious of continued claims about the existence and importance of human nature' (1986: 12). The biologist Michael Ghiselin has made the claim even more bluntly: 'What does evolution teach us about human nature? It teaches us that human nature is a superstition' (1997: 1). Their scepticism is not grounded in some special feature of humans; it is not, for example, an artefact of our capacity for culture and the variation that this capacity brings to our species. On Hull and Ghiselin's view, humans don't have natures because no biological species have natures: they are not the right sorts of things to have them. Could it be that the whole debate about interfering in, or improving on, human nature rests on an error?

---

[1] This chapter first appeared in *Philosophy and Technology* 25 (2012): 459–74. I am grateful to Luciano Floridi, two anonymous referees from *Philosophy and Technology*, and especially Russell Powell for incisive comments on an earlier draft. The research leading to these results has received funding from the European Research Council under the European Union's Seventh Framework Programme (FP7/2007-2013)/ERC Grant agreement no. 284123.

Cutting to the chase, the answer to this question is clearly 'no'. Hull and Ghiselin object to a rather particular picture of human nature, which one might call the 'essentialist' conception. Their arguments do not show that there is no respectable conception of human nature. Hull himself was quite explicit about this: 'If by "human nature" all one means is a trait which happens to be prevalent and important for the moment, then human nature surely exists' (1986: 9). In this chapter I explore the notions of human nature that are, and are not, biologically respectable, and I examine their uses and abuses in debates regarding enhancement. I begin by explaining the biological consensus, which is usually thought to make room for a minimal conception of human nature very much in line with Hull's own proposal. I go on to examine and criticize a more detailed elaboration of this conception, proposed by Edouard Machery (2008). It turns out that it is difficult even to sustain the minimal notion of human nature in the light of work on polymorphism and the role of learning in the development of widely distributed traits; the only biologically respectable notion of human nature that remains is an extremely permissive one that names the reliable dispositions of the human species as a whole. This conception offers no ethical guidance in debates over enhancement, and indeed it has the result that alterations to human nature have been commonplace in the history of our species. In the final sections, I examine a rather different, quasi-Aristotelian conception of human nature that is frequently invoked in ethical discussion, including discussion regarding enhancement. This conception is highly misleading. Moreover, because our folk psychology finds this misleading conception tempting, we are in fact better off if we refrain from mentioning human nature altogether in debates over enhancement.

4.2 The Biological Consensus

To the extent that there is any philosophical consensus regarding biological species, it is that biological species fall into an entirely different category of thing, metaphysically speaking, to chemical elements (see e.g. Sober 1980; Dupré 1981; Griffiths 1999; Ereshefsky 2001, 2008; Okasha 2002). An influential story about chemical elements has it that they are natural kinds with intrinsic essences. On this view, a lump of metal is properly classified as gold in virtue of its having the right microconstitution, and more specifically in virtue of its atoms having 79 protons. This microstructural property explains many of the characteristic macroscopic features of lumps of gold: their malleability, electrical and thermal conductivity, density, and so forth. This microstructural property also constitutes the sense in which all samples of gold are fundamentally the same: they all have the same atomic number. This also means that were a particular lump of metal to change its atomic number from 79, it would no longer be gold; it would belong to a different elemental kind.

One might think that a similar sort of story could be told about species: perhaps species, too, are natural kinds, again with microstructural essences. On this view, being a member of a given species is a matter of having the right genetic constitution. The genome explains the characteristic properties of each species, and it too constitutes the sense in which the members of a species are fundamentally the same. Once again, if the genetic constitution of an individual human, dog, or whatever is altered, it thereby stops being a member of that kind.

In fact, this way of thinking about species has been almost universally rejected among biologists and philosophers of biology. Most taxonomic philosophies instead regard species members as united not in virtue of possessing similar intrinsic properties, but instead in virtue of the relations they stand in to each other (Okasha 2002; Ereshefsky 2008). This stance is most clearly expressed by those who think of species as individuals, and not as kinds at all. It should be obvious that not all the entities we describe are natural kinds. The desk in my office is not a kind, it is an individual. An individual comes into existence at a given point in time (e.g. when the desk is built); it has a location in space which may change over time (as it moves from one office to another); it can survive considerable changes in its properties over time (as it is sanded, repainted, repaired); and it disappears at a moment in time (when it is destroyed). Importantly, the parts of a desk are not united because they are made of the same intrinsic materials: the legs may be metal, the top may be composed of wood and leather. Instead, they are united because of the relations of attachment they stand in to each other (Lewens 2007b).

Similar facts hold for species: they come into existence at a moment in time; they can change their location in space as their ranges expand or contract; they can survive changes to their properties over time as they are affected by selection, mutation, drift, and so forth; and they disappear at a moment in time (when the species becomes extinct). Individual organisms are parts of a given species not in virtue of the intrinsic properties they share, but rather in virtue of the relations they stand in to each other. What these relations may be is contested: some take it to be a matter of breeding relations, others take it to be a matter of the niche they occupy, for others it is a matter of ancestry (Okasha 2002; Ereshefsky 2008). The upshot of all this is that a given species can be genetically and morphologically heterogeneous at a given time; these genetic and morphological properties can change over time; and species membership is not constituted by the possession of some micro-essence. This does not mean that species have no essences at all: just as, on Kripke's view, individual entities such as tables and persons have their origins essentially, so one might argue that species, again understood as individuals, have their origins essentially (Okasha 2002). The key message from the consensus is not that species have no essences, but that species do not have explanatory micro-essences of the sort that would make them analogous to chemical kinds. That is why Hull and Ghiselin have objected to references to 'human nature' and to species natures in general.

This consensus is not, of course, without critics. Michael Devitt has recently argued that biological species do have intrinsic essences after all (Devitt 2008; 2010). Devitt's papers have attracted several rebuttals (e.g. Barker 2010; Ereshefsky 2010; Lewens 2012), and this is not the place to rehearse a response to him in full. Instead, we should note that even Devitt's version of essentialism is mild. His view is that internal genetic properties go some way to explaining the characteristic properties of biological species. For this reason, he thinks it appropriate to describe species as kinds, whose essences are partially intrinsic. Devitt's view, in other words, is that a causal explanation of (say) why tigers develop in such a way as to grow stripes can point, in part, to patterns of genetic material that tigers share. Having said this, Devitt explicitly concedes (a) that generalizations ranging over the members of a given species, such as 'tigers are stripy', are rarely (if ever) without exceptions, (b) that a full causal explanation for the reliable development of tiger stripes would have to point to non-genetic resources, and (c) that being a member of the tiger species is not a matter of having some single 'tiger gene'. Instead, Devitt takes it that some perhaps unruly pattern of genetic variation is characteristic of a species, hinting that species may be defined as loose 'genotypic clusters'.

Devitt's own brand of essentialism is committed to little more than the view that there are reasonably reliable generalizations about members of individual species, which are in turn partially explained by appealing to the genetic constitution of the species in question. These claims are not denied by proponents of the species-as-individuals view, for the mere fact that a species is properly regarded as an individual, whose parts are particular organisms, leaves open the question of how far natural selection, or other biological forces, may have acted to homogenize any given species, thereby giving rise to reliable generalizations about common traits within the species at any given time. While an individual need not be composed of materially homogeneous parts—and the example of the desk shows this—this is not to say that an individual cannot be homogeneous in some respects and at certain times. Since evolutionary forces can sometimes lead to the accumulation of difference and sometimes to homogenization, it is an open empirical question as to which of our psychological, behavioural, anatomical, or physiological features will turn out to be widely distributed across the species. Moreover, such distributions are vulnerable to change as the evolutionary forces acting on them change. That, to repeat, is why Hull explicitly endorsed a mild reading of 'human nature'.

4.3 Permissive Natures

Edouard Machery (2008) has recently tried to argue for a respectable notion of human nature, the detailed description of which he takes to be the goal of work in evolutionary psychology. Machery's conception of his project as a response to Hull is rather confusing, because much of what he says is in the spirit of Hull's view. Machery distinguishes an essentialist conception from what he calls a 'nomological conception', and he argues that the nomological conception is legitimate. Machery's nomological conception has it that 'human nature is the set of properties that humans tend to possess as a result of the evolution of their species' (2008: 323), and we have already seen that Hull did not deny that some properties are common among humans and that evolutionary processes are often responsible for this commonality.

We should be very clear at the outset that Machery's nomological conception is, as he himself recognizes, not morally normative: human nature consists in traits that are common among humans, but this does not mean there is anything morally wrong in failing to possess one of these traits, nor does it mean there is anything morally right in exemplifying the statistical norm. Like Hull's own tentative account of what human nature might be, Machery's proposal is merely statistical: human nature consists in nothing more than a set of traits that are widely distributed within the human species and which owe that distribution to any of a variety of evolutionary processes. Natural selection tends to promote the spread of those traits that augment an individual's capacity to reproduce, and yet there is no general reason to suppose that traits of this sort are ethically admirable. Moreover, natural selection is just one of a variety of evolutionary processes; in small populations a trait may become widespread as a result of drift. Such a trait may be less fit than alternatives. The upshot of all this is that on Machery's conception of human nature, the question of whether traits that are a part of our nature are also morally desirable is an open one. So Machery's conception of human nature would not serve as much of a foundation for someone trying to argue against (or for) modifications to the human species. Moreover, Machery's conception (like Hull's) underlines a point made elsewhere by Daniels (2009): to change human nature itself, rather than to change an individual so that he or she is no longer exemplary of human nature in some respect, would require changing the typical distribution of traits in the human population as a whole. On this view, to change human nature would be a tall order: it would require more than the development of a few individuals with extended attention spans or augmented athletic abilities.

Having said all this, there are some problems even with Machery's minimal account of what human nature is supposed to be. They are instructive when we address ethical issues regarding enhancement. There is a minor worry in the way he describes his project, for he tells us that: 'According to this [nomological] construal, describing human nature is thus equivalent to what ornithologists do when they characterize the typical properties of birds in bird fieldguides' (2008: 323). A field guide is designed to enable the correct recognition of an individual bird, qua member of a given species or subspecies, in the field. Since this is the function of field guides, it is not surprising that they restrict themselves to properties that are (a) diagnostic of a given species and (b) easy to observe over a short period of time (such as when out on a walk). Machery's conception allows that a property can be part of human nature when it is neither diagnostic of our species nor easy to observe; the property instead needs to be widely distributed through our species. Some such properties may be found widely among other species (hence they are not diagnostic), and they may only be apparent in an individual human after microscopic observation or prolonged observation. If evolutionary psychologists Daly and Wilson (1988) are correct, then adverse treatment of genetically unrelated offspring is a feature of human nature even though it is not the sort of feature one can easily observe on a short walk 'in the field', and it is not the sort of feature that is peculiar to humans: on both counts, it is not the sort of feature one would expect to find in the entry under 'Human' in a field guide to mammals. Much more generally, the vast number of homologies shared between humans and related species (pentadactyly, the possession of a spinal cord, and so forth) are parts of human nature on Machery's view even though they are not diagnostic of our species. It is misleading to say that the description of human nature under the nomological conception is equivalent to the sort of description one finds in an ornithological field guide.

That was merely a presentational problem; Machery's proposal has two more deeply problematic features. First, he argues that something is only a part of human nature when it is 'shared by most humans'; second, he argues that something is only a part of human nature when humans possess that feature 'as a result of their evolution'. Let us begin with the first requirement: 'male philandering' is just the sort of trait that evolutionary psychologists are likely to regard as part of human nature, and yet only half of humans are male (Buss 1999). Field guides, as Machery notes, typically contain several images for a single species, including representations of the species at different life stages, and representations of males and females. More generally, evolutionary processes including natural selection can often lead to various forms of polymorphism and sometimes produce very distinct morphs with characteristic anatomical and behavioural features. The marine crustacean species *Paracerceis sculpta* offers a nice example (Shuster 1987). Here there are three male forms: one large one that guards smaller females in harems contained inside sponges, one very small one that sneaks into the sponges, and one that imitates the females to enter the sponges in disguise. On Machery's account, a feature is only part of the nature of a species when it is shared by most of its members. What justifies this verdict, rather than various reasonable alternatives? Why, for example, shouldn't we be prepared to say that it is in the nature of the species *P. sculpta* to produce these three morphs, or perhaps that each morph has its own nature? Machery considers these problems and defends his understanding on purely pragmatic grounds: since other disciplines (he names forms of anthropology and personality psychology) focus on human differences, 'it is useful to have a notion that picks out the similarities between humans' (2008: 324). Even considered pragmatically, this justification is weak: after all, a conception of human nature that makes room for polymorphism need not overlook traits that are more widely distributed throughout the species. What's more, personality psychology and those forms of anthropology that highlight cultural difference are rarely in the business of offering evolutionary explanations. If Machery's aim in outlining a respectable concept of human nature is to ensure that an important realm of inquiry is not neglected, then the study of evolved polymorphisms would seem just as worthy of support as the study of traits that evolution has distributed more widely in our species.

This brings us to Machery's second major requirement for something to count as part of human nature: what does it mean to say that some feature of our species is 'a result of evolution'? This problem is compounded by his own insistence that 'nothing is said about the nature of the evolutionary processes in the proposed characterization of human nature. The traits that are part of human nature can be adaptations, by-products of adaptations, outcomes of developmental constraints, or neutral traits that have come to fixation by drift' (2008: 324). Of course this doesn't go very far towards explaining what does count as an evolutionary process, and here again Machery's definition takes the form of a stipulation: 'saying that a given property [...] belongs to human nature [...] is to reject any explanation to the effect that its occurrence is exclusively due to enculturation or to social learning' (2008: 326).

There are a number of ways in which we can put pressure on this stipulation. First, Machery's thought appears to be that social learning is not itself an evolutionary process; hence if social learning is the only process responsible for the occurrence of some trait, the trait does not count as a part of human nature. To make this argument good one needs some reason for denying that evolutionary processes in general include cultural evolutionary processes. In contrast to this, the likes of Richerson and Boyd (2005) have built fruitful evolutionary models that consider how populations change over time—how they evolve, that is—under the influence of various forms of social learning. Second, to the extent that 'human nature' is supposed to name those traits that are widely distributed among our species, and perhaps even those traits whose development is very robust or hard to evade, it is not clear why we should rule out an explanation for their persistence that looks solely to enculturation or social learning. We may think it unlikely that these processes would result in widely distributed traits, because we might think that where social learning is wholly responsible for the development of a trait it will also tend to result in considerable variability across different communities. This hunch is better left as an empirical claim to be established by proper investigation. Moreover, if it is true, then Machery's insistence that human nature only names traits that are characteristic of the species as a whole will ensure by itself that traits produced solely through these forms of learning will not feature in human nature.

Third, it is not clear that Machery's conditions succeed in excluding any widely distributed traits from human nature. He gives an example of a trait that he thinks is not part of human nature: 'the belief that water is wet is not part of human nature, in spite of being common, because this belief is not the result of some evolutionary process. Rather, people learn that water is wet' (2008: 327). Why should we not count this belief as the result of an evolutionary process? Machery suggests two rather different answers, but neither is satisfactory. The first response looks back to the notion that if a trait's occurrence 'is exclusively due to enculturation or to social learning', then it is not evolved. We have already seen reasons to question this stipulation, but Machery is quick to point out, 'This is of course not to deny that social learning, or indeed any other environmental influence, can be part of the explanation of the development of the trait' (2008: 328, fn 8). Once this concession is made, though, we will surely concede that since *part* of the development of my belief that water is wet refers to phenomena other than enculturation and social learning, such as my own perception of water's wetness, it can qualify as part of human nature after all.

Machery's second response is rather different: 'Saying that a trait has an evolutionary history is to say something stronger than the fact that it has perdured across generations. Humans have probably believed that water is wet for a very long time, although this belief has no evolutionary history. For this trait is not a modification of a distinct, more ancient trait' (2008: 327).

This may be a plausible claim for beliefs about water, but it looks less plausible if we consider a different example of a trait that is very widely distributed among the human species today, such as basic knowledge of the rules of association football. This knowledge has surely undergone modification over time, as the rules of the game have changed. If we follow Machery's second response, knowledge of the rules of association football is part of human nature, while knowledge of the wetness of water is not. Machery's proposal to equate human nature only with those elements of our common makeup that can be understood as modifications of earlier, more ancient traits threatens to draw a theoretically dubious distinction between different cognitive traits, all of which are produced via social learning.

These are not merely philosophers' quibbles: in recent work the psychologist Cecilia Heyes has argued that human social learning—that is, the ability to learn from other humans—in fact makes use of general learning abilities (the sorts of ability that underpin trial and error learning, for example), coupled to input mechanisms that are specifically tuned to the activities of other agents (Heyes 2012). Importantly, Heyes suggests that this tuning of input mechanisms—that is, the shaping of perceptual, attentional, and motivational systems that direct us to the activities of other people—may itself be influenced by further learning. Thus, Heyes considers the possibility that the motivation of children to copy the actions of others may itself be a learned product of the rewards adults give them for successful imitation (see Ray and Heyes 2011). The ability of humans to imitate others is just the sort of widely distributed trait that one would presumably want to count as part of human nature, and the question of the role of social learning or enculturation in its production is one that needs to be explored in full: it is not the role of a definition of human nature to exclude social learning or enculturation as the explanation (or, for Machery, the sole explanation) for the widespread development of some important feature of our species.

What lessons can we take away from our discussion of Machery's proposal for a respectable account of human nature? The answer is that an initially attractive conception slips through our fingers as we inspect it more closely. There is no particularly good theoretical or pragmatic reason to insist that the 'nature' of a species consists solely in those traits that are widely distributed throughout the species. This is especially clear when we focus on the conception of a biological species as an individual, that is, as a complex, structured entity in its own right. The human species, like all species, has various reliable tendencies, in virtue of the distribution of genetic and other developmental resources throughout it (Lewens 2009). It is in the nature of the human species to reliably produce organisms that are similar in certain respects, but it is also in the nature of the human species as a whole to reliably produce patterns of difference, such as the anatomical and physiological differences between the sexes (Ereshefsky and Matthen 2005). There is also no particularly good theoretical reason to insist that traits cannot count as parts of the species nature if they are solely produced by forms of enculturation or social learning; that is partly because it is unclear that any traits are produced *solely* through these channels, partly because learning processes can underpin the development of traits that are reliably distributed throughout the species, and partly because various forms of social learning appear to have been important in the evolutionary history of our species.

Having made these concessions, the nomological conception becomes disturbingly permissive, and it is hard to see how it can be reined in other than by unprincipled fiat. If we understand 'human nature' simply to name the reliable dispositions of the human species as a whole, then what grounds do we have to deny that it is in the nature of the human species to produce Catholics, or that it is in the nature of the human species to produce rugby players? These are certainly reliable tendencies of our species, which have endured over a reasonable length of time. We do not want to exclude them on the grounds that there have not always been Catholics or rugby players, because any biologically respectable notion of human nature must allow that it can change. Nor do we want to exclude them on the grounds that Catholicism or rugby-playing are learned, because any biologically respectable notion of human nature must allow that learning contributes to our makeup.

What's more, while changing the reliable dispositions of the complex entity that is *Homo sapiens* may be difficult, there will be a great many ways of intervening in the broad patterns of human development, all of which could potentially result in changes to human nature. Population-wide programmes of genetic manipulation may produce changes to human nature on this view, but population-wide changes to educational regimes may do the same. Even more restricted processes, such as imperialist interventions that extinguish certain languages or rituals, will count as alterations to human nature because they alter the reliable tendencies of the species as a whole. The alterations that have been going on in our species over the past few hundreds of years, wrought by the arrival of printing, the computer, and so forth, have certainly made broad differences to the ways in which humans develop; there is no good theoretical reason to deny that these amount to changes in human nature too. On this view we should deny Daniels's claim that the modification of human nature is exceptionally ambitious, for it is the sort of thing that is going on all the time.

Machery might reasonably point out that this more permissive account of human nature is so broad that no human trait seems to be excluded from its extension; this makes the account unacceptable. Instead, the permissive account must be reined in somehow, perhaps along the very lines Machery proposes. One might simply decree that 'human nature' is to name those properties that most humans share, and which are produced by a restrictive subset of evolutionary processes, even if there is no particularly strong theoretical reason to ground the chosen set of restrictions. The case for such a stipulation is, however, purely pragmatic: it does nothing more than to limit the field of inquiry for a particular empirical discipline in order to contain its subject matter. It is hard to see why one should wish to maintain this narrow stipulative understanding of 'human nature' when one is inquiring about the ethics of various interventions that might alter humans; here, the permissive account seems hard to resist. From this perspective it is hard to find fault in Marx's claim that 'all history is nothing but a continuous transformation of human nature'. And, of course, if human nature has been constantly transformed throughout history, it is hard to see how alterations to human nature could possibly mark out a type of intervention that will receive uniform ethical evaluation. Instead, ethical discussion of the rights and wrongs of diverse interventions in our species needs to focus on appraising the details of the interventions themselves, rather than the question of whether they might change 'human nature'.

4.4 Neo-Aristotelianism

In the last section I argued that the account of human nature we owe to Machery and Hull leaves open the question of whether there is anything morally wrong with altering human nature, and I also pointed to a variety of problems with that account, which all push us towards an exceptionally permissive view of what human nature is. Whether we accept Machery's restrictive account or move to the permissive conception of human nature, we fail to generate an account that offers any sort of ethical guidance. Of course, other writers in ethics and bioethics have implicitly or explicitly adopted very different conceptions of species natures, which they believe do have ethical implications, and I turn to these now. In this section I examine the appeals to human nature that underlie an influential form of neo-Aristotelian meta-ethics, championed recently by Michael Thompson (1995; 2008) and Philippa Foot (2001). I should explain at the outset that what follows is not intended as a refutation of their work, because I will not be addressing the most important parts of their case (although see Chapter 10 of this book for further criticism of Foot). Thompson, in particular, argues that reductive definitions of what it is to be a living thing all fail or beg the question, and he uses this as a clue to fashion his neo-Aristotelian account. This part of the argument falls beyond the scope of this essay. Instead, I simply want to focus on the support their work gets from the puzzling status of what Thompson calls 'Aristotelian categoricals'.

Thompson has drawn our attention to a peculiar sort of claim one encounters about species and their natures. People are prone to say things like 'The bobcat breeds in spring' or 'Man sheds his teeth' (these are Thompson's own examples). What are we to make of such statements, which Thompson calls 'natural historical judgements' or 'Aristotelian categoricals', and which take the form 'The S is (or has, or does) F'?

It is obvious that these statements shouldn't be understood as referring to a particular bobcat, or a particular man—they are instead efforts at more general description. Moreover, Thompson claims, quite plausibly, that we can't understand these claims simply as shorthand for 'Most bobcats breed in spring' or 'Most men shed their teeth'. He takes it that a statement like 'The mayfly breeds before dying' is not challenged by the fact that most mayflies die without breeding at all. Thompson consequently takes it that these sorts of statements, while describing a species, nonetheless specify standards for species members. This is borne out by the possibility that most species members might fail to meet the standards in question.

We will examine Thompson's positive proposal for how to understand natural historical judgements in a moment. First, let me quickly review some highlights from psychological and anthropological investigations of the 'folk' conception of species natures (e.g. Atran 1990; Atran et al. 1997; Gelman and Hirschfeld 1999). On the face of things, this empirical work gives considerable support to Thompson's and Foot's claims. We are, it seems, intuitive essentialists (see also Linquist et al. 2011). So, for example, research on young children has suggested that they regard each living kind as having some sort of underlying internal nature, which is itself causally responsible for the appearance of typical features of the kind in question. These essences are hidden: they can fail to manifest themselves properly even when they are present. They are teleological, in the sense that they are oriented to some end state that may not in fact appear. It would be wrong, then, to think that these statements about essence can be translated without loss of content into statements about statistical norms within the living kind, for the essence might continually 'misfire' and fail to result in an anticipated regularity. Finally, it has been argued that people conceive only of living kinds in this way. They do not regard artefacts or chemical elements as possessing internal 'teleoessences' of this type, for the making of forks is not driven from the inside of the fork and the making of carbon is not goal-directed. Atran et al. summarize: people 'presume the biological world to be partitioned at that rank [the generic-species rank] into non-overlapping kinds, each with its own unique causal essence, or inherent underlying nature, the visible products of which may or may not be readily perceived' (1997: 39). The end result of all this is that the folk do indeed seem committed to the existence of 'natures' for (folk) species, which can be described, which are irreducible to statistical generalizations about individual organisms, and which inhere solely in the organic world. These essences also have normative implications, in the sense that they specify proper and improper developmental outcomes in virtue of their teleological orientation. (This does not entail, of course, that species natures are conceived of as entailing fully ethical norms for species makeup.) Such thinking emerges very early in cognitive development; as Gelman and Hirschfeld put it, 'Four-year-old children act like essentialists, assuming that members of a category share an innate potential and that innate potential can overcome a powerful environment' (1999: 418).

Although they do not cite this ethnobiological literature, both Thompson and Foot use these features of folk discourse to strengthen their case for a particular view regarding meta-ethics. In both cases they take it that there are natural facts, of a rather special sort, which ground claims about proper standards for species functioning. Thompson describes the project he and Foot share: 'the suggestion was made that practical reason be viewed as a faculty, akin to the powers of sight and hearing and memory; it was further maintained that an individual instance of any of these latter powers is to be judged as defective or sound by reference to its bearer's *species*' (Thompson 1995: 250). This is a sort of neo-Aristotelian position, according to which we can learn about the right way for individual members of a species to be by learning about the nature of that species. Needless to say, this is the sort of meta-ethical position that might have ramifications for the way in which one goes about evaluating alterations to individuals that cause them to depart from the proper nature of their species. Whether it also has implications for proposals to change human nature is unclear, because even if one thought that the nature of a species specified normative standards for its members, it still would not be clear whether altering that nature (and hence altering the standards) would be problematic.

The problem, of course, is that while empirical research supports the notion that the folk are committed to the existence of species-specific teleo-essences, such research does nothing to show that species really have such essences. The psychological researchers in question have taken it as a given that while people might be tempted to think of species in this way, there is no justification in biological reality for that conception (Gelman and Hirschfeld 1999: 405). The onus is on the likes of Foot and Thompson to show not merely that much discourse seems committed to normative essences, but that there really are such essences. Unless such a task is achieved, they have shown only that people are intuitive Aristotelians, without showing that Aristotelianism is a position we should adopt. The confusion regarding what implications their position might have for attempts to alter human nature is, of course, an artefact of a further problem with their underlying meta-ethical position: with no clear explanation of what makes it the case that a species' nature is one way rather than another, there is also no clear answer to the question of when an intervention changes the nature of the species, and when it simply changes the observed properties of individual members of the species while leaving its underlying (and potentially hidden) nature intact.

Another way of posing this problem draws once again on empirical research. The folk do indeed take it that there are internal essences for each species. However, they also seem to hold that there are internal essences for other living categories, including sexes and races (see Gelman and Hirschfeld 1999 for references and discussion). This reminds us that biological species are not the only things that can play the role of subjects in Thompson's natural historical judgements. Darwin, for example, was prone to mention 'the Negro' or 'the Australian' (1877/2004: 184), and he was not trying to pick out a particular Negro or a particular Australian. Darwin thought of the Negro as 'light hearted, talkative', in contrast with the 'taciturn, even morose' aborigine of South America (1877/2004: 198). In *The Descent of Man* Darwin quotes with approval William Greg's description of '[t]he careless, squalid, unaspiring Irishman' (1877/2004: 164), but that should not move us to think that 'the Irishman' is a successfully referring term, even if Darwin and Greg thought it was. We should not object to Thompson's claims that 'a concept is a species concept if it is a possible subject of the corresponding form of judgement [i.e. a natural historical judgement]. A life-form or "species" (in the broad sense) is anything that is, or could be, immediately designated by a species concept or life-form word' (Thompson 1995: 292). The psychological evidence suggests that we do indeed make use of species concepts, or life-form words, in Thompson's sense. This does not mean, however, that there are any species, or life forms, so understood: 'the Irishman', 'the Negro', and even 'the mayfly' fail to refer.

One might reply by saying that it is not only 'the folk' who think of species in this way; they are conceptualized like this in bona fide biological documents too. Thompson, for example, thinks of field guides as describing, either implicitly or explicitly, the Aristotelian natures of species. Surely we do not wish to deny the legitimacy of field guides?

Thompson may be right about the interpretation of field guides, but once again this does not help to show that there are any Aristotelian natures so understood. First, recall that the function of a field guide is to enable recognition. Many ethnobiological researchers have argued that our intuitive essentialism facilitates various forms of recognition and classification—it allows us to rationalize the appearance of otherwise troublesome variation, for example, and may sometimes summarize central diagnostic features—even though they regard these essences as having no reality in nature. It would be no surprise, then, if something like an Aristotelian essence were depicted in a field guide. Moreover, recent research by Linquist et al. (2011) reminds us that even scientists can continue to be subject to the cognitive biases of folk essentialism; we should not expect such misleading conceptions to be stamped out in works used by scientists. Finally, a field guide is not a technical document summarizing the genuine features of a given species and detailing the range of potential variation that might occur among its members; its images serve heuristic functions that are used either by amateurs or as a precursor to the more technical application of a species key. So we cannot point to the fact that species natures appear in field guides to argue that they must correspond with real entities.

4.5 'Our Given Nature'

Leon Kass is usually understood to be a recent champion of another form of neo-Aristotelian thinking about species, whereby their natures also specify ethical standards for their members. This is the way that Buchanan (2009) has interpreted him. It is also the way Adam Briggle has understood things in his book-length study (2010) of the work of the US President's Council on Bioethics, which Kass presided over. Briggle takes himself to be describing and endorsing Kass's own view when he tells us, in neo-Aristotelian fashion, 'The nature of *Danaus plexippus* [i.e. the monarch butterfly] is seamlessly descriptive and normative, as it defines what constitutes full flourishing within the pattern of that kind' (2010: 91).

It is odd, then, to see how often Kass's own writings seem explicitly to repudiate this sort of neo-Aristotelian view. My primary aim in this penultimate section is to expose this confusing character of Kass's own appeals to human nature, and in so doing to underscore and elaborate a point made elsewhere by Buchanan (2009), and made in a preliminary form at the end of Section 4.3: we would be better off leaving human nature out of ethical discussion altogether.

In a discussion of Michael Sandel's work on human enhancement, Kass himself tells us that the mere fact that some feature is a part of human nature leaves its ethical status open. He puts it like this:

In short, only if there is a human givenness, or a given humanness, that is also good and worth respecting, either as we find it or as it could be perfected without ceasing to be itself, does the 'given' serve as a positive guide for choosing what to alter and what to leave alone. Only if there is something precious in the given—beyond the mere fact of its giftedness—does what is given serve as a source of restraint against efforts that would degrade it. (Kass 2003: 20)

Kass is here formulating an objection to Sandel's celebration of the 'giftedness' of human nature. Human nature is 'gifted' in the sense of being a 'given': it is something we find humans to have. Kass makes the compelling point that the mere fact that something is part of human nature tells us nothing about whether it should be celebrated or exterminated (see also Chapter 2). Perhaps some elements of human nature are indeed worth preserving and celebrating, but if so we need to establish that they are 'precious' and to explain why we should preserve them: 'The mere "giftedness" of things cannot tell us which gifts are to be accepted as is, which are to be improved through use or training, which are to be housebroken through self-command or medication, and which opposed like the plague' (Kass 2003: 19).

Kass's discussions in this paper and elsewhere are puzzling, because they sit uneasily between the unobjectionable practice of building a moral case for celebrating certain features that are widely distributed among humans and the objectionable practice of using the fact that some feature is part of human nature to give a moral justification for it. When he tells us that 'cloning shows itself to be a major alteration, indeed a major violation, of our given nature as embodied, gendered, and engendering beings' (Kass 1998: 689), is he claiming that cloning will damage an aspect of human nature that he can also show to be valuable, or is he claiming that cloning is objectionable because it constitutes an alteration of human nature?

Consider Kass's claim that 'Sexual reproduction [. . .] is established [. . .] by nature; it is the natural way of all mammalian reproduction' (1998: 689–90). Perhaps all Kass means by this is that, as a matter of fact, mammals reproduce sexually. But then what is the function of insisting that this has been established 'by nature'? On the face of things this is a redundant claim: what else other than 'nature' could we possibly think has established sexual reproduction? The contrast he is drawing here is between what is established 'by nature' and what is instead established 'by human decision, culture, or tradition' (p. 689). This is a spurious contrast: we have already seen that there are no good reasons to think that culture and tradition are not parts of nature, and that they may not be involved in the generation of widely distributed human traits. As Kass himself says, 'There can, in truth, be no such thing as the *full* escape from the grip of our own nature. To pretend otherwise is indeed a form of hubristic and dangerous self-delusion' (2003: 18). We cannot escape from nature because whatever we do—even if that means augmenting our cognitive abilities with drugs or reproducing through asexual cloning—is partly determined by our prior goals, which are themselves ultimately produced by natural processes. The fact that Kass feels it necessary to say not merely that 'mammals reproduce sexually' but that this is 'the natural way' suggests he thinks there is some additional force, perhaps ethical force, in describing something as 'natural'.

Kass's comments on Sandel remind us, though, that further ethical reasoning is required to show whether sexual reproduction should be 'improved', 'housebroken', 'opposed like the plague', and so forth. To his credit, this further reasoning is just what Kass attempts to provide us with. He explains that a colleague asked him whether he would have opposed efforts to move away from asexual reproduction, had this instead been the reproductive system established in humans. It is telling that Kass's response departs, as it should, on a discussion of why he believes that sexual reproduction is worth celebrating. The implication seems to be that had we been asexual 'by nature', Kass would have welcomed efforts to transform us into sexual beings. Being sexual is, Kass thinks, morally richer than being asexual: 'For a sexual being, the world is no longer an indifferent and largely homogeneous otherness [. . .] It also contains some very special and related and complementary beings, of the same kind but of the opposite sex, toward whom one reaches out with special interest and intensity' (1998: 691). It is unlikely that sexual reproduction is truly an important factor here: plants are sexual beings, but this does not suffice to enrich the moral character of their interbotanical lives. Kass attempts to back his position up with the claim that 'it is impossible [. . .] for there to have been human life—or even higher forms of animal life—in the absence of sexuality and sexual reproduction'. Many evolutionary biologists will agree that the ability of sexual reproduction to foster variability has indeed been a contributor to increased complexity in the plant and animal kingdoms, with the result that sexual reproduction has indeed been instrumental in producing the higher faculties that we value. It is another thing to say that these faculties would not survive if some human reproduction became asexual.

We will come on to these substantive issues in a moment; the important point to note here is that Kass's basic argument seems to be that sexual reproduction is worth preserving not because it is 'natural', but because its erosion would lead us to regard the world in the way asexual beings do, namely, as 'an indifferent and largely homogeneous otherness'.

Kass continues to make the case for the poverty of asexual reproduction by telling us that we 'find asexual reproduction only in the lowest forms of life: bacteria, algae, fungi, some lower invertebrates' (Kass 1998: 691). This isn't true: parthenogenesis—a form of asexual reproduction whereby female eggs develop without fertilization from males—is regularly observed in vertebrates, especially reptiles. But even if Kass were right, it is hard to see what conclusions we should draw for human relationships from the sadly impoverished lives of asexual bacteria. Sexual reproduction in a given human couple hardly seems necessary for them to harbour deep feelings for each other, and it is not necessary to maintain the other-regarding attitudes that may accompany their raising children. Many humans forgo sexual reproduction to raise adopted children, and for some of them when they 'reach out with special interest and intensity' it is to 'special and related and complementary beings' of the same sex. Suppose, then, that a clone is formed when the nuclear material from one woman is inserted into the enucleated egg of another woman she loves, and the resulting embryo is implanted into the second woman's womb. This is 'asexual reproduction' in the technical sense that the nuclear genetic material comes from one parent only, but there are still two partners with an essential biological involvement in the generation of new life, there are still genetic contributions from both parties (for the gestational mother contributes her mitochondrial DNA), and there may still be plenty of 'reaching out' to each other, both when the decision is made to begin a family and later as the new person's life unfolds. If Kass is concerned to preserve a world in which rich relationships flourish between other-regarding humans, more argument is needed to show that asexual reproduction will undermine it.

4.6 Beware of 'Human Nature'

The tendency to linger in debates over cloning and enhancement on what is 'natural' to humans should be abandoned (see also Buchanan 2009). It either constitutes an irrelevant preamble to the important question of which features of human reproduction should be preserved, or it constitutes an objectionable allusion to some mythical and morally loaded 'human nature' that might serve as an ethical yardstick in debates of this sort. What is more, the fact that we seem to be intuitive Aristotelians about species natures in general should make us particularly wary of the resonances of framing debates about enhancement in terms of claims about human nature. Our tendency is not merely to understand claims about human nature as casual statistical truths about human populations, with no particular ethical significance, but instead as descriptions of an underlying, normatively laden, internal essence we all share. This reinforces the basic message of David Hull's work: Hull allowed that there might be a respectable notion of human nature, but he was rightly troubled when human nature was put to work in ethical and political debate.

5

From *Bricolage* to BioBricks™

Synthetic Biology and Rational Design

5.1 Engineering Nature

General accounts of the nature of synthetic biology have systematically stressed that it involves using principles of rational design for the fabrication of organic systems.[1] This aspect of the project is repeated in a slew of definitions of synthetic biology. Here are just a few, drawn from the UK and European contexts. The UK Health and Safety executive says that synthetic biology can be described as 'the design and construction of new biological parts, devices and systems, and the redesign of existing, natural biological systems for useful purposes' (HSE 2007). The European Commission says that synthetic biology is 'the engineering of biology: the synthesis of complex, biologically based (or inspired) systems which display functions that do not exist in nature. This engineering perspective may be applied at all levels of the hierarchy of biological structures [...] In essence, synthetic biology will enable the design of "biological systems" in a rational and systematic way' (European Commission 2005). The UK Parliamentary Office of Science and

---

[1] This chapter was first published in *Studies in History and Philosophy of Biological and Biomedical Sciences* (2013). An early, and very different, version was presented at the *Organism and Machine* meeting in Copenhagen in January 2011. I am grateful to Russell Powell and Sune Holm for organizing that meeting, and to the audience for many valuable questions. I also owe thanks to Beth Hannon for supplying information on the aurochs case, and to Staffan Müller-Wille, Maureen O'Malley, and John Dupré for generous comments on recent drafts of the paper. The research leading to these results has received funding from the European Research Council under the European Union's Seventh Framework Programme (FP7/2007-2013)/ERC Grant agreement no. 284123.

Technology tells us that synthetic biology 'describes research that combines biology with the principles of engineering to design and build standardised, interchangeable biological DNA building-blocks. These have specific functions and can be joined to create engineered biological parts, systems and, potentially, organisms. [Synthetic biology] may also involve modifying naturally occurring genomes [...] to make new systems or by using them in new contexts' (POST 2008).

There are several significant differences between the sub-varieties of synthetic biology (O'Malley et al. 2007). Some approaches aim at the 'bottom-up' construction of genetically specified components and at their combination for the creation of novel organic devices. Others instead begin with naturally occurring organisms and seek to remove redundant or unnecessary parts in order to produce simpler, less 'noisy' organic machines, or perhaps minimally functioning organisms. In spite of markedly different attitudes towards the centrality of the genome in organic functioning, and the detailed methods for the production of novel biological systems, more recent research articles maintain the notion that an engineering design perspective characterizes a variety of synthetic biology approaches. So, for example, synthetic biology has been described as 'the use of engineering techniques to model, design, and construct artificial biomolecular networks' (Camacho and Collins 2009: 24), and even more recently we hear again that at 'the heart of synthetic biology lies the goal of rationally engineering a complete biological system to achieve a specific objective, such as bioremediation and synthesis of a valuable drug, chemical, or biofuel molecule' (Cobb et al. 2012: 1).

What is the significance of this talk of rational design methods? Is this approach entirely new or merely a variant on a long-standing theme within biology? Does it signal some ethically troubling attitude to the natural world, perhaps drawing on the 'impulse to mastery' that Michael Sandel (in a different context, examined in more detail in Chapter 2) has claimed characterizes various efforts at human enhancement (Sandel 2007)? I begin in Sections 5.2 and 5.3 by arguing that while humans have influenced organic lineages in many ways, and in spite of the fact that some of these have involved the intentional manipulation of organic traits for human ends, it is nonetheless reasonable to place synthetic biology towards one end of a continuum between purely 'blind' processes of organic modification at one extreme and wholly rational, design-led processes at the other. In Section 5.4 I use an example from evolutionary electronics to explore some of the constraints imposed by the rational design methodology itself. These constraints reinforce the limitations of the synthetic biology ideal—limitations that are often freely acknowledged by synthetic biology's own practitioners. In Section 5.5 I conclude by arguing that the synthetic biology methodology reflects a series of constraints imposed on finite human designers who wish, as far as is practicable, to communicate with each other and to intervene in nature in reasonably targeted and well-understood ways. This is better understood as indicative of an underlying awareness of human limitations, rather than as expressive of an objectionable impulse to mastery.

5.2 Designing Nature

A first attempt to articulate the notion that there might be something new about synthetic biology's engineering approach could draw on the pretensions of that school to take control of natural processes and subject organic nature to intentional manipulation. A moment's reflection shows that in this respect at least, synthetic biology is simply an extension of an ancient tradition whereby humans influence the makeup of the organic world. Even if synthetic biology threatens to blur the distinction between organisms and artefacts by constructing an organism, that distinction has never been exclusive. A dairy cow is an organism if anything is. And yet, dairy cows have clearly been modified by human breeders with the purpose of increased milk yield in mind. Dairy cows are organisms that have been purposefully manipulated: they are organisms and artefacts at the same time. Artificial organisms have been around for as long as intentional agents have practised artificial selection. Indeed, even before humans consciously practised artificial selection, their symbiotic relationships with plants and animals have exerted mutual evolutionary pressures on our own lineage (e.g. as lactose tolerance followed the domestication of dairy cows) and on other lineages (e.g. through the domestication of the wild ancestors of dogs).

Organisms, then, have long been artefacts, too; the enlarged udders of the modern cow owe their enormous volume to the efforts of generations of humans. But the pretensions of synthetic biology are not merely to influence the organic world in line with human goals: as we have seen, synthetic biologists wish to design organic objects using rational methods. We might stop short of claiming that dairy cows have been 'designed'. Why? To say that something has been designed is to say not merely that some of its modified features are intended or desired outcomes, but that its structure has been planned. The process of design typically involves drawing up, or conceiving of, *a design*. It involves considering likely uses of the imagined artefact and laying out an articulated structure for that artefact in the light of these potential uses, perhaps in the form of a diagrammatic specification, in advance of its construction (Cross 2000; Houkes and Vermaas 2010). Pioneering breeders of the 18th century like Robert Bakewell developed significant technical expertise regarding the most effective techniques of artificial selection. Bakewell stressed the importance of strong selection and to some extent he pioneered the use of 'progeny testing'—that is, the evaluation of a parent's ability to transmit desired traits to offspring via an assessment of a significant number of those offspring (Russell 1986). In the case of both sheep and cattle, Bakewell aimed at the creation of a profitable meat animal, and he even went so far as to formulate rather more specific targets for how this should be achieved: animals should grow quickly, they should transform fodder into meat in the most efficient way, the animal's mass should be given over as much as possible to saleable meat, and that meat should be of the most economically valuable kind. This adds up to a design specification for the intended outcome of the breeding project only in a highly informal sense. Nicholas Russell has argued that while some of Bakewell's stated goals may have been quite detailed, his actual methods were less refined: 'There can be no doubt that phenotypic conformation must have formed the basis of his selection programme. He chose animals which looked right, which in his terms meant those that were easy to fat, were thin-legged (since that was the place to observe the bone size in the live sheep) and conformed to the shape which he believed would reflect the best carcass for flesh and fat distribution' (1986: 201). This suggests that synthetic biology really does offer something new, not because it blurs the organism/artefact boundary—that has always been blurry—but because of its goal of bringing the organic within the realm of design, where design is understood to carry all the connotations of planning, diagrammatic representation of the device to be constructed, standardization of parts to be assembled, and so forth that feature in the engineering design process.

It is quite rare to find an explicit discussion of what makes a design process rational, as opposed to irrational, but alleged exemplars of the two processes can be found with ease. In pharmaceutical discovery, 'irrational' processes include the use of vast 'libraries' of combinatorially generated molecules, coupled to high-throughput screening methods that find molecules with the desired functions. 'Rational' processes, by contrast, include efforts to construct valuable molecules using well-understood chemical principles, where the chemical knowledge in question both explains and predicts the desired functional outcome. 'Irrational' methods can be perfectly rational, in the sense that they often constitute sensible strategies for the production of an entity with the desired overall function. They are 'irrational' only because the agent who oversees the process need not understand why the resultant entity works as it does. Whether a method is 'rational' or 'irrational' is itself something that comes by degrees; an engineer may not understand why the parts of a machine work as they do, but he may still be able to predict in rough-and-ready ways how they will behave when brought together. After several moderately successful attempts he may be surprised at the makeup of the machine that works best. Here, his knowledge is partial, and some elements of irrationality are introduced into the process. Likewise, an animal breeder may have some rudimentary understanding of a desired physiological function, and he may have a good sense for which animals it is best to mate together. He is not quite in the situation of one who merely collects together a vast array of animals and picks the ones that suit his purposes best; at the same time, he falls well short of assembling parts whose combined function he predicts with forethought and accuracy.

Synthetic biology represents an effort to introduce rational design methods, in the sense that synthetic biologists attempt to produce parts with stable functions, which can then be combined in ways whose outcomes are reasonably transparent to rational prediction. We must be careful, however, not to exaggerate how sharp the break is between synthetic biology and what has gone before. I will give just two examples, the first from the early Mendelians, the second from a series of more recent breeding programmes. Some early Mendelians, in a manner evocative of some of synthetic biology's recent proponents, had a vision of animal and plant breeding as a practical endeavour whose precision would be greatly increased by knowledge of genetics (Müller-Wille 2012).

They believed that ultimately it would be possible to exert fine-grained control of the characteristics of organisms via the combinatorial manipulation of allelomorphs. William Bateson put forward such a view in an address to the New York Horticultural Society in 1902, which was published two years later. The links drawn between this proto-synthetic biology and synthetic chemistry, and the vision of rational control replacing the breeder's intuitive knowledge and luck, make it worth reproducing the quotation in full:[2]

> To use an illustration: In chemistry you may have a body, say, a simple salt, from which you can take out the base, or the acid radical, replacing the base by another base, or the acid radical by another acid radical. You can in that way decompose your substance into component parts, reforming them in various combinations. So we must imagine a plant which has one element of color, for example, another element of texture, etc., and we must conceive that when two varieties are crossed together the unit characters can be combined and recombined in the gametes of the hybrid, alternating with and replacing each other by substitution. You can take out greenness and put in yellowness; you can take out hairiness and put in smoothness; you can take out tallness and put in dwarfness, etc. The characters have their fixed possibilities of union, and hence it may be possible for us to form some mental picture of the constitution of the organism.
>
> Now when we come to the question of the significance of these things to the breeder or to the hybridist, it will be found that the significance is exceedingly great. I am afraid of saying that we have already reached a point when the practical man who is doing these things with a definite, economic object or commercial object in view can take the facts and use them for his definite advantage. But we do for the first time get a clear sight of some of the fundamentals on which he will in future work, and it cannot be now very many years, if the investigations go on at the present rate, before the breeder will be in a position not so very different from that in which the chemist is: when he will be able to do what he wants to do, instead of merely what happens to turn up. Hitherto I think it is not too much to say that the results of hybridization had given a hopeless entanglement of contradictory results. We did not know what to expect. We crossed two things; we saw the incomprehensible diversity that comes in the second generation; we did not know how to reason about it, how to appreciate it, or what it meant. We got contradictory results, and the thing looked hopeless. But with the discovery of the purity of the germ cells we have the first step, which,

---

[2] A scholarly detail: Bateson's remarks are often significantly misquoted, because a short paraphrase suggested by Levins and Lewontin (1985: 180) has been understood by their readers as a direct quotation. The quotation has also been misattributed, because Levins and Lewontin erroneously reference the (1902) 1st edn of Bateson's *Mendel's Principles of Heredity* as the source of the comment.

I think, is bound in a very short time to become a path through many of those wonderful mazes of heredity. (Bateson 1904: 2–3)

In summary, then, the aspiration to modularized, rational control over the design of artificial organisms, one of the defining themes of synthetic biology, was already present in the earliest years of genetics. Moving to slightly more recent breeding programmes, there have long been efforts, even before the advent of synthetic biology, to draw up plans for intended organisms, and then to seek to bring these design specifications to reality, albeit through traditional programmes of selective breeding rather than through the fabrication and juxtaposition of standardized organic elements. Some of the best-known programmes of these sorts include efforts to reconstitute the now extinct aurochs. The programme begun by Heinz and Lutz Heck in the 1920s, which subsequently enjoyed considerable support under the Nazis through Heinz Heck's friendship with Hermann Göring (Millet 2011), is today viewed as an unsuccessful and simplistic attempt to 'breed back' an extinct species. Heinz Heck's later comments nonetheless express an ambition to combine hereditary elements in a modular manner with the aim of reconstituting the aurochs piece by piece, in a fashion consonant with Bateson's earlier remarks:

No animal is ever utterly exterminated as long as some of its hereditary factors remain. The fact that these qualities may not be visible is shown by the laws of heredity to be unimportant, for what is hidden may be brought to light again and, by cross-breeding, the original component parts may be isolated. In the case of the Aurochs conditions are favourable since all its physical characteristics are still present and to be seen. They are, of course, divided between many different breeds of cattle, one having preserved a good aurochs horn, another its build, while a third has the characteristic colouring, and so on. (Heck 1951: 119–20)

Commentators today tend to regard these cattle as very unlike the aurochs, even if they may have some superficial morphological similarities. Van Vuure's verdict is damning:

On the basis of vague criteria and without proper knowledge of the appearance of the aurochs, the two brothers made inaccurate selections among the crossbreeding products of various cattle breeds. They did not use the knowledge about the aurochs that was available at the time, nor did they take advantage of the breeding techniques others were using to create new cattle breeds in the same period. (Van Vuure 2005: 366)

In spite of this apparent failure, what are now called 'Heck cattle' are still used in conservation efforts around Europe. The more recent—indeed ongoing—*TaurOs* project (http://www.taurosproject.com/) is presented as a more elaborate (and commercially oriented) attempt to first ascertain the likely genome of the aurochs, to discover its ecological setting, to reassemble an aurochs-like breed by combining genetic elements from suitable extant cattle, and to insert the resultant animal into an appropriately structured environment (Faris 2010). The *TaurOs* project involves a more formal process of design, in the sense that it begins with a fine-grained specification for the desired genomic and ecological profile of the target breed, in spite of the fact that its breeding techniques are those of more traditional artificial selection. Whether this project will be any more successful than that of the Hecks is an open question.

We can now see more generally that there is a continuum between 'pure' forms of natural evolution at one extreme, where no intentional agency influences the evolution of a lineage, and efforts at fully formalized and standardized rational design processes at the other. Lying between these extremes we have the forms of unconscious selection that Darwin (1859; 1868) pointed to, whereby merely because humans choose to preserve some varieties of cereal, say, they unconsciously end up modifying their stock. We have intentional efforts to modify domesticated breeds for the better, even when the principles of breeding are not well understood. We have the kind of sophisticated and targeted efforts at improving particular traits (meat quality, milk yield, and so forth) that characterized the emergence of many famed agricultural breeds during the 18th century. We have the clumsy design efforts of the Heck brothers and genomically assisted endeavours like the *TaurOs* project. Synthetic biology merely lies further towards the 'design' end of that continuum than anything we have witnessed before.

5.3 Creativity and *Bricolage*

We have established that there is a 'design continuum' and that synthetic biology lies towards one end of it. We will shortly see that, having acknowledged the existence of a continuum of this sort, it is possible to misplace synthetic biology at a far extreme of that continuum, by exaggerating the degree to which synthetic biology eschews more traditional methods of artificial selection and by exaggerating the efficacy of its efforts at pure rational design. The positioning of synthetic biology towards one end of this continuum is frequently cited as a differentiating feature between it and genetic engineering (e.g. Endy 2005). Where genetic engineering typically aims at the piecemeal modification of existing organisms via changes to the genome, synthetic biology instead aims (in some of its guises at least) either at the wholesale creation of organisms or at the creation from scratch of a series of organic machines, via the assembly of modularized sub-parts. The aspiration to produce a set of standardized sub-units that can be combined in reliable and predictable ways perhaps reaches its clearest expression in the BioBrick™ programme. Here, an 'open-source' registry of standardized genetic parts is maintained, and the BioBrick™ parts are available for use in, among other things, the annual iGEM competition, where teams submit their new biological systems. This move to a fully design-based approach is sometimes claimed not merely to be a novel transition in terms of the ambition we have to control nature, but a transition with moral weight. Boldt and Müller (2008: 387–8), for example, have recently asserted: 'The shift from genetic engineering's "manipulatio" to synthetic biology's "creatio" is a shift with considerable ethical significance.'

I will return to Boldt and Müller's comment in more detail a little later. For now, note that it seems to rely on a fairly strong distinction between mere manipulation—perhaps understood as a form of piecemeal 'tinkering' with what nature happens to present to us—and full-blown creation. This is a distinction we should be sceptical of. For example, there is a sense in which the process of natural selection itself is both supremely creative—for it produces novel adaptations where none were present before—and supremely constrained to tinker with the materials that natural variation happens to make available to it. In describing evolution as a form of 'bricolage', François Jacob clearly drew an analogy between the blind process of selection and the creative human *bricoleur* who fiddles with bits and pieces available to him, at the same time as he highlighted the suboptimal outputs we should expect from selective processes. Darwin (1868) tried to illustrate the creativity of natural selection itself by sketching the image of an architect who judiciously chooses and assembles rocks of diverse shapes that have fallen from a precipice; the creativity of the process is not undermined by the lack of control over the availability of the raw materials. Indeed, many theorists of engineering creativity itself have tended, in turn, to conceptualize technical innovation and technical progress in evolutionary terms, describing both in a vocabulary that stresses analogies between invention and natural selection (e.g. Vincenti 1990; Ziman 2000). Historians of technology and philosophers of science, too, have been attracted to an evolutionary image of scientific and technical creativity, according to which novel artefacts, or novel theories, consist in suitably reconfigured and modified combinations of elements inherited from their historical ancestors (e.g. Basalla 1988). On such views, creation is nothing more than a series of fortunate instances of manipulation, extended over a long enough time period to produce something that seems genuinely novel compared with its ancestors.

Finally, and most significantly for Boldt and Müller's thesis, while synthetic biology might have the *ideal* of producing a series of elementary functional units that can be combined in ways that produce larger wholes with predictable and valuable functions of their own, in *practice* this remains merely an ideal. Kwok (2010) gives a sobering list of challenges facing the BioBrick™ methodology. Practitioners of synthetic biology repeatedly stress the practical need for later phases of directed evolution, whereby variants are produced using mutagens, and tested for the desired overall function, if a suitably performing molecular machine is to be constructed (see O'Malley 2011 for further exploration of this theme). This emphasis on combinations of rational and irrational methods has persisted over several years of methodological work in synthetic biology (see e.g. Andrianantoandro et al. 2006; Dougherty and Arnold 2009; Porcar 2010; Cobb et al. 2012). As it is practised right now, synthetic biology does not manage to avoid an essential phase in which rational design methods are put aside, in favour of a process of artificial selection of artificially generated variants. In other words, the creativity of synthetic biology itself continues to rely on processes whose efficacy may be inscrutable to the designers themselves, and which have much in common with familiar breeding techniques. In this respect, synthetic biology still has one foot in earlier traditions of artificial selection, and many recent commentators have suggested that it might always remain in this situation. In an article advocating the use of techniques of 'directed evolution' for the de-bugging and optimization of rationally designed components, Dougherty and Arnold begin by remarking:

[Some] researchers advocate efforts to 'standardize' biological parts in such a way that their behavior in novel assemblies or environments becomes more predictable. The notorious complexity and context-dependency of the behavior of biological parts and their assemblies, however, make such standardization extremely challenging. It is unlikely, in fact, that biological parts can ever be fully standardized, and engineering methods that enable rapid optimization of synthetic biological systems will be very useful. (Dougherty and Arnold 2009: 486–91)

The tools of directed evolution are here described, reasonably enough, as 'engineering methods', but that should not obscure the fact that these methods do not embody rational design principles as traditionally understood. As currently practised, perhaps as it will always be practised, synthetic biology should not be placed at a far extreme of our 'design continuum'. The proper design continuum should, therefore, look something like that pictured in Fig. 5.1. Crucially, given the complexities of biological systems, the far right extreme of that continuum may never be reached, but may instead have to serve as an aspirational goal that directs and spurs research.

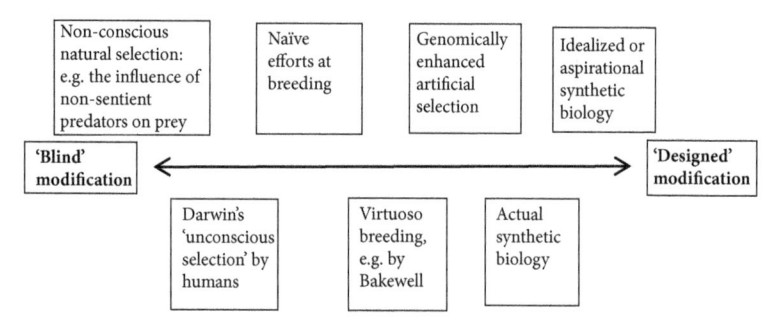

Non-conscious natural selection: e.g. the influence of non-sentient predators on prey	Naïve efforts at breeding	Genomically enhanced artificial selection	Idealized or aspirational synthetic biology

'Blind' modification ←――――――――――――――――――――――――――→ 'Designed' modification

	Darwin's 'unconscious' selection' by humans	Virtuoso breeding, e.g. by Bakewell	Actual synthetic biology	

Fig. 5.1 A design continuum for the modification of organic nature

5.4 Evolutionary Electronics

There is no clear sense, then, in which the application of design methods characteristic of synthetic biology represents an entirely new way of approaching the organic world: at best it is a further development in an ongoing trend. But we can begin to see some of the more interesting features of rational design methods if we take the unusual step of looking at what happens when innovators explicitly throw away rational design principles and instead use 'blind' methods of innovation similar to natural selection. It turns out that the sorts of designs that are produced are often very different to those thrown up by rational design. Indeed, some of the proponents of these evolutionary design processes claim that rational design processes—the very processes that advocates of synthetic biology wish to apply to the organic realm—can act as strong constraints preventing the exploration of significant areas of design space. If this is so, we might temper our enthusiasm (and perhaps our fear, too) for the likely ability of synthetic biology to radically re-order the biological world.

Synthetic biologists often explicitly frame their goals by analogy with electronics. They look to create organic logic gates, for example, that can be combined to produce larger components—oscillators, amplifiers, input and output devices, and so forth—in much the same way that electronic engineers can assemble electronic components. It is especially interesting, then, to consider what happens when people working on electrical devices themselves throw away attempts to design components in a rational way. Consider an example from evolutionary electronics, specifically from Adrian Thompson's group at the University of Sussex in the UK. The following case is explained in detail in Thompson (1997). The group used something called a field programmable gate array (FPGA)—essentially a reconfigurable chip—whose organization was determined by a genetic algorithm. They tested a series of slightly modified configurations to the chip, preserving the circuits that were best able to discriminate between two audio tones. Although the FPGA had a 64×64 matrix configuration, the programmed section of it used only a 10×10 area of one corner of the chip. After thousands of iterations, a version of the chip that performed the task more or less perfectly was bred. The exact details of the investigation do not matter here. But the final piece of hardware had some extraordinary properties. Most striking for our purposes is the fact that some sections of the best-performing evolved chip could not be altered without loss of function, even though they were not electrically connected (in the usual way, at least) to the output of the chip:

[These cells] cannot be clamped [i.e. returned to their default 'blank' state] without degrading performance, *even though there is no connected path by which they could influence the output* [...] They must be influencing the rest of the circuit by some means other than the normal cell-to-cell wires: this probably takes the form of a very localised interaction with immediately neighbouring components. Possible mechanisms include interaction through the power-supply wiring, or electromagnetic coupling. (Thompson 1997: 399–401)

These comments also demonstrate the important point, common in evolutionary design approaches, that when a high-performing variant is bred, investigators are frequently unable to understand why it works as well as it does. I do not wish to assert that it would be impossible for the operation of such circuits to be understood, and we have just read of some of the Sussex group's suggestions for how this might be achieved. The point, instead, is that because unconstrained pathways of evolutionary design are free to take advantage of whatever fitness-enhancing effects may present themselves, there is no reason to anticipate that such effects are easy for observers of the system to recover and understand.

We should not be surprised that evolutionary processes can give rise to entities that are both highly functional and incomprehensible. In processes of rational design we begin with a general problem and divide it into sub-problems. We then construct elements that will solve each of these sub-problems, and we stick them together. It is essential, in order for this method to work, that the functional capabilities of the sub-parts are robust across contexts, in the sense that they don't change their internal configurations when they are brought into the vicinity of other parts, that their interactions with other functional units are always predictable, and so forth. In other words, we try to ensure that sub-parts are causally isolated from each other, and from the specifics of their environments, as far as is possible. We aim to construct modularized components. This is not because causal isolation is necessarily a virtue in terms of the overall desired effect; it is simply because the design methodology of decomposition into sub-problems demands this sort of causal isolation. The end result of all this is that the operation of designed entities is comparatively easy to recover and understand through a functional decomposition of the object into constituent parts with contributory effects. This is precisely because the systems have been designed, and their parts have been assigned discrete functions during the design process itself. While such comprehensibility is likely to be a feature of designed systems, it is not guaranteed even in the realm of artefacts. Even here innovators may have unwittingly failed to reduce forms of cross-talk that, unknown to them, make an important contribution to the overall operation of the artefact.

Evolution, on the other hand, cannot literally divide a problem into sub-problems, to be tackled in isolation (Lewens 2004a). This also means that when fortuitously beneficial aspects of 'cross-talk' between elements of a system are hit upon in evolving systems, they will often be retained. There is certainly no literal effort to remove them. This result has costs and benefits. In the case of Thompson's reconfigurable chip, it turned out that when their high-performing 10×10 configuration was moved to a different part of the larger 64×64 chip, the performance was impaired or even dropped off altogether. So costs, in terms of portability, are often extreme. But, as the Sussex group rightly notes, this means that evolutionary design techniques can sometimes be considerably better than standard rational design techniques, precisely because a variety of potentially beneficial effects that rely on 'cross-talk' between elements, or contingent properties of the locale, can be harnessed, rather than minimized as a result of the design methodology itself:

> Evolution has been free to explore the full repertoire of behaviours available from the silicon resources provided, even being able to exploit the subtle interactions between adjacent components that are not directly connected. (Thompson 1997: 401)

This form of evolutionary process unsurprisingly gives rise to systems whose overall functioning is often inscrutable to human investigators. Where no effort is made to reduce forms of 'cross-talk', to minimize the ways in which overall functioning is contingent on highly idiosyncratic properties of local material substrates, and so forth, it can become very difficult to subject these systems to typical forms of functional decomposition. These work best when components have discrete functions, whose joint actions are simple additive results of their independent contributions. In this sense, then, we can understand some of the key differences between paradigm cases of designed machines and the much messier workings of evolved systems—whether those systems are carbon-based organisms or inorganic circuits. John Dupré, in recent work, is right to stress the sense in which the creation of a highly modular, well-behaved system that is apt for unambiguous functional decomposition is itself a significant achievement (Dupré 2012). It is the sort of thing that humans strive to produce via modular design methodologies.

5.5 Rational Design and Evolutionary Design

Reflection on these evolved electronic systems is instructive when we begin to think about synthetic biology. Even in the biological realm, theorists have claimed significant advantages for modularity. But it is important to understand why modularity is held to be important: an evolutionary system that was very tightly connected, so the story goes, would be unable to evolve in a cumulative manner. Suppose that any modification to the eye, or to the heart, would tend to have knock-on effects on every other organ in the body. The chances are that any potential positive impact on the design of the eye would be counteracted by the countless impacts—most likely negative—on all of the connected elements of the organism. And so, the story continues, modularity is essential for cumulative complex adaptation.

Modularity may indeed be important, but on this view its importance does not lie in natural selection's methodological practice of decomposing a problem into sub-problems. As such, even if modularity is a frequent feature of organic design, we should not expect modularity to be a hard-and-fast feature of all evolutionary solutions. If it turns out that non-modular designs are sometimes effective, then selection knows no better than to choose them. Moreover, while theorists have often stressed the ways in which partial modularity characterizes many organic systems, it remains unclear whether natural selection is able to favour modularity. Some have argued that the emergence of modularity is better explained as a result of non-adaptive processes (e.g. Lynch 2007), while some more recent models have proposed possible routes whereby selection for modularity might proceed (e.g. Pavlicev 2011).

This discussion enables us to note some important rules of thumb which distinguish most organisms from most artefacts. Most artefacts will have been built in such a way that their layout reflects a prior conceptualization, by its own architects, of the structure of the design problem the artefact addresses. This is an observation often made by engineering design theorists, and even by management theorists with an interest in innovation (Cross 2000; MacCormack et al. 2008). It has frequently been conjectured (and some evidence supports the conjecture) that the structure of a complex designed product typically 'mirrors' the organizational structure of the corporation that produced it (MacCormack et al. 2008). That is because the overall problem addressed by the artefact needs to be broken down into sub-problems of a sort that can be tackled by independent teams. The availability of such teams, and their particular expertise, is constrained in turn by the organizational composition of the company they work for.

Needless to say, there is no analogue to this phenomenon in nature. Even when organisms are modular, this sort of conceptualization is evidently not the cause of that modularity, and in many cases departures from modularity will be favoured by natural selection. This, for example, is why distinct organic *functional* modules—that is, complexes of organic processes that combine to perform valuable developmental or physiological roles—need not be *structurally* isolated from each other, and their elements can often overlap. If a given biomolecule can be turned to useful roles in numerous different functional processes, then so be it.

Where have we got to in our discussion? There really is something rather novel in applying engineering principles to the natural world. An organism—or a sub-organic machine—organized according to the principles of rational engineering design would be far more modular, far more compartmentalized, than any natural organism. Such an organism will feature clear efforts to reduce 'cross-talk' between its functional elements. This does not mean that we should expect artificial organisms to be superior to natural organisms in all aspects of their performance: in many cases there are advantages to be had from harnessing these forms of cross-talk, as the examples from evolutionary electronics indicate.

But now recall why we should care about standardization: synthetic biology's advocates typically do not argue that the reason to move to some form of rational design approach is because such an approach inevitably arrives at outcomes superior to those produced by natural selection. We have already seen some reasons for thinking that in restricted cases natural selection may produce more effective designs than rational agents. Natural selection is not constrained by a prior methodological commitment to rationality, and since natural selection must build traits that are effective in the face of regular internal and external insults, natural selection also needs to ensure that its products are robust against a range of environmental perturbations. The forms of redundancy and complexity that follow from blind design may often be superior in performance to the products of rational methods. Synthetic biology's advocates more usually adopt the pragmatic premise that if humans are to make efforts to alter what nature has given us, or to build new organisms, we are likely to wish to do so in ways that allow conversations between designers, the organization of design teams, the swapping and transportation of effective elements from one design context to another, and so forth. This is an attitude one quite frequently encounters in publications within synthetic biology. For example, a very recent paper notes that many biological functions one may wish to harness from a technological point of view are 'encoded in gene clusters', with the result that their naturally occurring genetic substrates exhibit various forms of complexity and inscrutability that make them very difficult to modify or improve upon:

> Regulation is highly redundant [...] regulation can also be internal to genes [...] Further, genes often physically overlap, and regions of DNA can have multiple functions. The redundancy and extent of this regulation makes it difficult to manipulate a gene cluster to break its control by native environmental stimuli, optimize its function, or transfer it between organisms. (Temme et al. 2012: 7085)

A more 'bottom-up' approach that makes use of modular synthetic components is then proposed, precisely because this offers a more tractable way to go about modifying and improving functions of biotechnological interest. These sorts of simplifications may also be of use in understanding biological complexity, because they represent a first attempt to understand simple systems, which may be overlaid one onto another in more biologically realistic, complex evolved systems (Morange 2009). Biological complexity is here acknowledged, the barriers it presents to human intervention are explicitly admitted, and attempts are made to avoid it (in the first instance, at least) via the creation of simpler systems.

Recall that Boldt and Müller (2008) have claimed that this shift 'from "manipulatio" to "creatio ex existendo" is decisive because it involves a fundamental change in our way of approaching nature'. On my reading, synthetic biology simply recognizes diverse practical reasons for aiming at various forms of modularity, standardization, and specification of subproblems. It is hard, then, to accuse synthetic biology of espousing an objectionable form of mastery over the feeble efforts of Mother Nature, of a wish to 'play God', or of some kind of Promethean assault on the 'given' of the natural world. These are the sorts of criticisms laid at the door of enhancement technologies by the likes of Michael Sandel (Sandel 2007; see also Chapter 2); some commentators might have similar concerns about synthetic biology. In reality, synthetic biologists simply recognize the practical constraints imposed by the methodology of rational design. Humans *cannot* play God, because God—unlike humans—is not constrained in these ways. God, one assumes, does not need to decompose problems into sub-problems; he only makes systems modular when a modular solution is the best one; he takes advantage of all forms of noise and cross-talk when they are there to be found; he has no need to make his plans comprehensible to finite human engineers. If God is omniscient, he simply surveys all possible solutions to a design problem and picks the best one, regardless of how complex it might be, regardless of how little modularity its elements may show, regardless of how hard it might be to understand its operations.

Maureen O'Malley (2011) has recently stressed, in a manner echoed by the comments in Section 5.3, that practitioners of synthetic biology have not been able to simply put together novel biological systems from functionally isolated parts and have done with it. Instead, there is always a 'kludging' or 'debugging' phase: rational design does not suffice, because biological systems demonstrate various forms of contingency, uncertainty, and context-specificity, which means that functional elements that work in one location cannot be transported without loss of function into another. Often their newly designed systems only work after considerable amounts of trial-and-error rejigging. If synthetic biologists were really trying to 'play God', the very existence of this final phase would be an embarrassment or admission of defeat. But on O'Malley's analysis it is simply acknowledged by synthetic biology's own practitioners as a reality to be faced when dealing with the contingencies of biological systems:

> All of this heterogeneity and evolutionary innovation has consequences for the type of engineering that can be done in synthetic biology. Diverse sources of variability obstruct synthetic biologists from achieving the desired 'plug and play' of predictable properties. Even when it works, rational design requires multiple iterations of reconstruction and redesign. Combinatorial synthesis and directed evolution—both employing 'irrational' biological processes to improve the functioning of designed devices—are increasingly necessary complements to or even replacements of rational design. (O'Malley 2011: 410)

She cites numerous biological sources for this assertion, including this interesting comment from Arkin and Fletcher:

[...] unlike other engineering disciplines, synthetic biology has not developed to the point where there are scalable and reliable approaches to finding solutions. Instead, the emerging applications are most often kludges that work, but only as individual special cases. They are solutions selected for being fast and cheap and, as a result, they are only somewhat in control [...] (Arkin and Fletcher 2006: 4)

Of course, none of this means that there are no significant ethical problems that arise when we look at synthetic biology. We need to examine who shall have control of the technology and to what ends. Some have worried that synthetic biology presents special challenges here because of the low costs of entry and because of its potential to be translated into biological weapons or instruments of terror. We need to examine how funding should be allocated to research in the area. We need to ask all the usual questions about how to balance uncertain prospects of benefit against uncertain prospects of harm. Some have worried that synthetic biology is particularly challenging because it does not fit easily into the standard safety paradigm derived from genetic modification, whereby modified organisms are compared with their natural counterparts. General ethical approaches to synthetic biology have looked at issues relating to biosafety and biosecurity, to concerns around intellectual property, issues relating to governance, and the 'exclusion' of important voices in the formulation of regulatory policies. These are serious issues, but they represent familiar ethical concerns about how to approach emerging technologies.

Synthetic biology involves something new, but only in the sense that it involves a further step towards one end of the 'design continuum': it involves an effort to apply rational design principles to the organic world. But when we understand why one might want to make that effort, and what its potential limitations are likely to be, we see that there is no special reason to worry that synthetic biology's practitioners are trying to fly too close to the sun. Synthetic biology, in spite of some of the slogans attached to it, does not represent an intolerable hubris.

6
Origins, Parents, and Non-identity

6.1 Origin Essentialism

In a review of a book about Saul Kripke for the *London Review of Books*, Jerry Fodor complains about the silly problems philosophers have become preoccupied with:[1]

> For example: 'I have never crossed the Himalayas, though I might have done. So there is a non-actual [...] possible world [...] in which someone crosses some mountains. Is that person me, and are those mountains the Himalayas?' [...Could] that really be the sort of thing that philosophy is about? Is that a way for grown-ups to spend their time? (Fodor 2004: 17)

A related problem, with a Kripkean aroma, has also detained bioethicists. They have become convinced that, had the circumstances under which my parents conceived me been somewhat different, the resulting child would not have been me. Moreover, they have drawn significant normative conclusions from these thoughts about the fragility of identity. David McCarthy, for example, writes:

> Suppose that a child is born as the result of medically assisted sex selection and has a disease or disability which was caused by the process of sex selection but nevertheless has a life worth living. Had that process not been used, the child would almost certainly not have come into existence at all. It would have been almost certain that either another sperm would have fertilised the egg or the egg would not have been fertilised at all. In either case, the child would not exist. It is therefore implausible that the child has a complaint since it has a life worth living and had sex selection not been used, it would not have existed at all. (McCarthy 2001: 304)

---

[1] This chapter is published here for the first time. I am grateful to Arif Ahmed and Christopher Stephens for extremely valuable comments on a very early draft.

McCarthy's argument seems to rely on the premise that a given individual could not—or more likely would not—have existed if a different sperm had fertilized the egg from which it sprang. I share Fodor's discomfort in addressing these speculative questions about whether an individual exists under various alternative possible circumstances; but if we want to assess a form of argument that appears frequently in bioethical circles we have little choice but to plunge into the metaphysical depths. What are the plausible modal constraints on an individual's reproductive origins?

Kripke and others (e.g. Kripke 1980; Forbes 1980; Salmon 1981) have argued, in one way or another, that the origins of an object are essential properties of that object. The considerations used to support this claim are usually wholly generic—that is, they do not turn on what kind of object is under consideration (Salmon 1981: 199). If objects have their origins necessarily, and persons are objects, then persons have their origins necessarily, too. The claim that persons have their parents necessarily is sometimes put forward as an obvious variant of a more generic origin essentialism. So Kripke (1980: 113) tells us that 'what is harder to imagine is [the Queen] being born of different parents. It seems to me that anything coming from a different origin would not be this object.'

In this chapter I am not interested in the question of whether origin essentialism is true or false. What I want to show is that origin essentialism only yields the proposition that persons have their parents necessarily, and the proposition that persons have their original gametes necessarily, in conjunction with some contestable premises about the nature of development and the role of genes in it. I will not evaluate those further premises in any detail. The point of this chapter is simply to show that the move from generic origin essentialism to gamete essentialism or parental essentialism is not as straightforward as Kripke's comments above might lead us to think.

The basic gist of my argument can be seen if we think of an analogous case: the origins of cakes. When we look at the ways in which interpreters of Kripke have sought to defend origin essentialism, we see that they are all keen to preserve the intuition that the origins of an object could have been somewhat different from what they are (see Robertson 1998 for references). So suppose I make a cake by adding sugar, then butter, then eggs to a bowl and mixing them. It is one thing to say that this very same cake could not have originated from a completely different set of ingredients. It is something much stronger to say that this very cake could not have originated from the same blob of butter, the very same eggs, but a dose of sugar taken from a different packet. Origin essentialism, as it has been advanced and defended, has not supported the strong claim that none of an object's origins could have been different.

So if origin essentialism allows that a small proportion of an object's original materials could have been different, it would also seem to allow that a small proportion of a person's original materials could have been different. Since there is a lot more to a person's origins than his original gametes, origin essentialism appears consistent with the denial of *gamete essentialism*—the proposition that a person could not have come from different gametes—and consistent also with the denial of *parental essentialism*—the proposition that a person could not have had different parents. Of course it does not follow from the claim that *some* portion of an object's origins could have been different that *any* portion of an object's origins could have been different—perhaps I could have come from a different placenta, but not from different gametes—but if we are to say that gametes have this special status among original resources, then we need some principle that tells us why this is so. In brief, some further premise is needed to get from origin essentialism to both gamete essentialism and parental essentialism. At the end of the chapter I will explain why this result has importance for invocations of the 'non-identity problem' in bioethics.

6.2 Focusing the Debate

Having sketched the argument, let me flesh things out a little more. One set of problems needs to be put to one side for scrutiny later, when we consider the conception of genes as information bearers. We will all agree, I suppose, that a token cake could have been made using a set of ingredients, some proportion of which differs from those used in the actual world. But what proportion of the cake's origins can change with the preservation of identity? And supposing we give an answer (say, a maximum of 50 per cent), what units should we use to count these proportions? Shall we say two cakes can be identical only if at most 50 per cent of their origins are identical by mass, or by volume? Perhaps some of the ingredients will be more important than others in determining identity according to their role in generating the distinctive properties of the final baked object. These questions are difficult to answer, but the important proposition to keep hold of at this stage is that origin essentialism is implausibly strong if it asserts that if *any* of an object's original materials had been different, then that object could not have existed.

A second set of problems concerns how to demarcate original resources that become parts of an object from those that are merely causally efficacious in its origination. Constructing an organism and constructing a cake both rely on the provision of a fairly rich range of materials, supplied at the right times. Far more goes into the original construction of an organism than the gametes of the two parents. Biologist Patrick Bateson first used the cake analogy to illustrate the mistake one makes in thinking that adult form is prefigured in genes (see Bateson and Martin 1999). Successful human development to adulthood will require, among other things, an appropriate set of nutritional resources, both inside the womb and outside; a properly functioning womb; suitable clothing and shelter, and so forth. Not all of these things become parts of the organism, and similarly not all the resources required to bake a cake (an oven, for example) become parts of the cake. My argument will assume less that is controversial if I restrict my attention to those things that do become parts of the adult organism. The crucial (and obvious) point to note is that the material resources that are incorporated over time into the developing embryo exceed merely the male and female gametes.

A third set of problems concerns how we will understand what a parent is. Near the beginning of his short discussion of origin essentialism Kripke asks us to assume that 'parents are the people whose body tissues are sources of the biological sperm and egg' (Kripke 1980: 112). This means that a world in which the embryo from which I actually developed was instead transferred to the womb of a woman other than my actual biological mother does not count as a world where I have different parents: parents are defined as gamete suppliers, and the gamete suppliers in this alternative world are the same as in the actual world. I will go along with the definition of parent as 'gamete donor' for the purposes of this chapter. With this definition in mind, the question for gamete essentialism should be understood as the question of whether an individual could have originated from different gametes. And the question for parental essentialism should be understood as that of whether an individual could have come from gametes which themselves came from persons other than the individual's actual parents.

A fourth set of problems arises when we try to extend arguments about origin essentialism to persons. The problems arise because we have little idea which ingredients supplied to a growing person should count as elements of that person's origins and which should count as bringing about transformations of the person after origination is complete. The question has relevance to the issues of gamete and parental essentialism. If we are very restrictive on what we count as origins, and include only resources packed into the gametes and perhaps also the extra-gametic environment at conception, then the gametes will form a very high proportion of original resources, and it is more likely that origin essentialism will reckon gametes essential to the individual. If, on the other hand, we are very relaxed about what we count as origination, and we don't think of a person's origination as complete until he or she is an adult, then—in material terms at least—the gametes will constitute only a very small fraction of the original materials of the person. I think a fairly liberal view is probably the right one: a cake hasn't been made when one has put just a handful of ingredients into the bowl, and a person hasn't been made when only the gametes have met. Even so I will assume, so as not to load things too much in my favour, a fairly restrictive notion of origins in this chapter.

6.3 Gamete Essentialism

With these preliminaries out of the way, here is an example that shows why gamete essentialism is not entailed by origin essentialism. Suppose that Alice and Ben conceive, producing baby Charlie. Charlie comes from a zygote, which we will name $Charlie$. $Charlie$ is formed from two gametes—call them $Alice$ and Ben. Consider alternative circumstances where Ben is not the spermatozoon that fertilized $Alice$, but instead a different gamete from Ben, which we can call $Ben*$. Imagine that neither Ben nor $Ben*$ houses genes that confer major pathology. Imagine also that $Ben*$ fertilizes $Alice$ at the same time as Ben fertilizes $Alice$ in the actual world. We can now see that origin essentialism requires some rather specific additional premises before we can conclude that the individual resulting from the union of $Alice$ and $Ben*$ is not Charlie. The origins of Charlie are different in this possible world, but only some of them. Most of the items that go into the production of the resulting individual are identical in this world and the actual world. Although Ben is missing, $Alice$

is still present, as is the environment of Alice's womb. On liberal views of origins, the subsequent nutritional and social resources later supplied by both Alice and Ben can be included, too. We would not want to deny that I could have made the same cake with a different cup of sugar. Why, then, deny that Charlie could have been produced from $_{Ben^*}$, rather than $_{Ben}$?

Perhaps the analogy with cakes is inappropriate, because while $_{Ben}$ surely makes up only a tiny proportion of Charlie's original materials by mass or by volume—consider how much larger an egg is than a spermatozoon—we might think that there is something especially important for Charlie's identity in the gamete that comes from his father. Maybe we do think this, but this alleged importance of gametes for identity cannot be justified by origin essentialism, which (as I have observed) is a wholly generic view about the essential properties of objects. Someone who wants to defend this move will need to independently motivate acceptance of the premise that gametes are important in fixing identity, without appealing merely to the fact that they are among the original materials of a person.

One obvious response to this challenge is to say that original materials are precisely that—materials from which an object *first* arises—so because an organism originates in the fusion of two gametes, each one of those gametes comprises half of the original materials. That, in turn, would motivate the further claim that a world where even one of the gametes is different from the actual world is a world where a very high proportion of original materials is different, so origin essentialism does support gamete essentialism after all. But once again, this response appears to make origin essentialism an implausibly strong view. It makes the very first element that goes into an object's construction an essential property of that object's history. Suppose my cake (call it Casper) is made by first placing a cup of sugar (call this quantum of sugar 'Suzy') into a bowl, and then a further cup of sugar (call this quantum of sugar 'Simon'), then butter ('Boris'), then eggs ('Egbert'). On the reading of origin essentialism that makes the very *first* item that goes into the construction of an object essential to it, Casper essentially has Suzy as the first of its original constituents. But we have already seen that origin essentialists want to preserve the intuition that some of an object's original materials could have been different, and on the face of it there is no reason to deny that the very earliest of its original materials could have been different, so long as the later original materials are the same. Consider the world in which I make a cake in the same circumstances as Casper, but where the timing of original materials is changed so that I put Simon in the bowl first, then Suzy, then Boris, then Egbert. Intuition certainly sides with this cake being Casper, and origin essentialists would surely be keen to preserve that intuition. So the fact that gametes are supplied very early in the material construction of a person does not ease the entailment of gamete essentialism by origin essentialism.

Now, while origin essentialism does not *entail* gamete essentialism or parental essentialism, it is perfectly compatible with both of those views. The basic plausibility of origin essentialism as a generic metaphysical doctrine requires that some small proportion of an object's original materials could have been different in circumstances where that same object still exists. In the context of persons, this requires at a minimum that, if we consider some actual person, then that same person could have existed even if some of the molecules from which she originated were different. But to say that there are worlds where this is the case is not to say that there are worlds where that person has different gametes, for these different molecules may make up the same gametes in all these worlds. This response shows how it is *consistent* to uphold a weaker form of origin essentialism, which allows that some of an object's origins could have been different, at the same time as endorsing gamete essentialism; but still we have not seen an argument for why we should believe gamete essentialism.

6.4 Parental Essentialism

Is origin essentialism consistent also with the denial of parental essentialism? Here is a case where the origin essentialist might agree that an individual exists in spite of neither of his actual parents being the source of the gametes from which he comes. Suppose in the actual world Amy and Brian have a baby called Cressida. Now consider a close world where Amy and Brian are both infertile. They use a donated egg and donated sperm to create an embryo, which is implanted into Amy's womb at the same time as (in the actual world) Cressida is conceived. As the embryo develops it goes on to receive the same nutrients in Amy's womb as the actual Cressida, the same baby food as the actual Cressida, the same schooling and cultural stimuli as the actual Cressida, and so forth. Why would we deny that this alternative child is Cressida? Although she differs from the actual Cressida in terms of the first two cells that produced her, she nonetheless shares with Cressida a huge array of other original resources supplied after conception.

Of course this example—like almost all examples regarding identity across worlds—trades on some highly questionable intuitions, and some may deny that such a child is Cressida. Perhaps a more convincing example, where at least one parent of an actual person is different, can be found in the close worlds where the nucleus is removed from the oocyte that forms the actual Cressida, and replaced with a complete set of nuclear material from some donor. Perhaps Amy asks her sister to provide the genetic material. The technology in question might approximate to maternal spindle transfer—a technique currently suggested as a means by which mitochondrial disorders might be eliminated, and which I described in Chapter 1. The oocyte is then fertilized with the very same spermatozoon from Brian that gave rise to the actual Cressida, and is reimplanted into Amy's womb at just the moment the actual Cressida was conceived. Once again, the developing cells receive the same nutrition in the womb, the same nutrition after birth, and so forth as the actual Cressida. In such a world we have a situation in which the resulting child shares no nuclear genetic material with its actual mother, yet the child does share a very high proportion of its non-nuclear original materials with Cressida, including the extra-nuclear materials packed into Amy's oocyte (mitochondrial DNA, cell membranes, and so forth).

Kripke's definition of a parent is inadequate to this case, for while Amy remains the source of the egg that produces the child, and is in this sense a parent of the child, Amy's sister is also a source of biological material that goes into the original formation of the child, hence also a parent, and she is the source of all this child's nuclear genetic material. It is quite plausible to say in these worlds that the child is Cressida, yet Amy does not donate nuclear genetic material to Cressida. Hence, Cressida could have had a different nuclear genetic mother, even on a view that regards origins as essential properties of objects.

Even this example may fail to convince. Many readers are likely to persist in the view that this person is not Cressida, or at least they may think that we are at a loss to say decisively whether this person is Cressida. But what the example does show, again, is that origin essentialism itself does not provide any straightforward support for the intuition that this person is not Cressida. Some other principle is needed in order for those who hold that intuition to justify it.

6.5 Informational Genes

Kripkean origin essentialism, by itself and in its generic form, supports neither gamete essentialism nor parental essentialism. But why have people tended to assume that these are both straightforward consequences of origin essentialism? It must be that some further hidden premise is widely accepted. One possible candidate for that job is the claim that genes are the most *important* of original materials, so for that reason we should give them particularly high weighting when assessing identity across worlds. This response tells us that we made a mistake in thinking of a world where an individual has different gametes as analogous to a world where a cake is made with a different cup of sugar. Genes, this argument goes, are not just more ingredients that go into the growing child; they have special significance among those ingredients. One might defend this claim by adding further premises that clarify how origin essentialism should be understood in the context of persons. So one might answer the first set of problems I raised in this paper for the formulation of origin essentialism by saying that for persons, origin essentialism is the claim that the original *information* that specifies the construction of a person could not have been significantly different. And one might then add that it is because genes constitute the bulk of information that produces an individual, that a person's genome could not have been significantly different. Further steps would then try to show, first, that this means a person could not have originated from different gametes (or perhaps that they could only have originated from different token gametes with near-identical genetic sequences) and, second, that this means that a person could not have had different parents, except in the most outlandish of possible circumstances in which different parents produce near-identical genotypes. It seems to me that anyone who upholds parental essentialism as a consequence of origin essentialism will need to defend premises resembling those suggested in this paragraph.

This claim about what is required for a defence of parental and gamete essentialism is a significant result in its own right. This is not a chapter on the informational conception of the gene, but I can add here a list of reasons why such a defence is likely to be a difficult job. Philosophers of biology of various persuasions (e.g. Griffiths and Gray 1994; Sterelny et al. 1996; Oyama 2000a, 2000b) have objected to singling genes out as the sole, or even the most important, bearers of developmental information on the following grounds, many of which are further elaborated in Chapter 7:

a) Genes do not suffice for the creation of an organism.
b) The possession of some gene does not guarantee that an organism will develop some phenotype.
c) Genes are not the sole developmental resources that have been selected for their role in guiding development.
d) Genetic and environmental factors can all be said to make predictable differences to developmental outcomes in specified contexts, and in this sense they all have a claim to the status of bearers of developmental information.

More recently, some commentators have made persuasive arguments for viewing the genome as a bearer of information, in a sense that does not immediately apply to all causal contributors to development. But even in these rather special senses of 'information', theorists have tended to resist the thought that the genome is the *only* bearer of information. Nicholas Shea, for example, has put forward a well-worked-out vision of evolutionary information, according to which only *inheritance systems* can count as information bearers (Shea 2007, 2012, 2013). His account of what makes something an inheritance system requires a background of natural selection for the function of assuring cross-generational resemblance. Yet Shea explicitly denies that the genome is the sole bearer of information in this sense; so long as we think forms of imitative learning have also been selected for the bringing about of parent–offspring resemblance, we should say that here, too, we find a source of information affecting the developing phenotype.

What is more, when other theorists have attempted to defend a sense of information that would indeed restrict it to the genome, they typically do so in a manner that is wholly unsuitable for someone who wishes to argue that the genome has special importance for the determination of identity. Peter Godfrey-Smith (2007), for example, has pointed out that we can reasonably argue that triplets of nucleotides *code* for amino acids by virtue of the well-known and highly stable mapping relations between the former and the latter. This sort of very tight and near-universal mapping relation does apply to the relationship between triplets and amino acids, but it does not apply to the relationship between (for example) alternative nutritional regimes and morphological phenotypes, even if the former causally affect the latter. Godfrey-Smith is right about all this, but the asymmetry between genome and environment he draws our attention to is of no use to someone who wishes to argue for the general developmental importance of genes. The problem is that while triplets have a code-like relationship with amino acids, the causal influence of genes on phenotypic traits that are further downstream in development—the sort of morphological, behavioural, or cognitive traits that we are likely to think central to assessing identity—is just as messy and sensitive to local context as the causal influence of nutritional environments on those same traits.

6.6 Bioethical Consequences

Why is the denial of an entailment of parental essentialism and gamete essentialism by origin essentialism important? As we saw at the beginning of this chapter, it seems that inferences based on the apparent consequences of origin essentialism sometimes underlie fairly common patterns of reasoning in bioethics. In a useful and insightful article on the ethics of modifying the mitochondrial genome, Bredenoord et al. ask whether 'germ-line modification leads to the birth of a different person'. They suggest that:

> In the case of other reproductive techniques, such as preimplantation genetic diagnosis, one chooses one embryo over another for implantation. If one decides to transfer embryo A instead of embryo B, this clearly leads to the birth of a different person. In the case of germ-line modification, this is less clear [. . . G]erm-line modification is likely to affect the qualitative identity of the future person, whereas his/her numerical identity will be untouched. (2011: 98)

These authors are suggesting that alternative reproductive technologies might be assessed in very different ways, in spite of their very similar purposes. Compare two hypothetical couples, both of which aim to have a child who is deaf. In the first case, they make use of what Bredenoord et al. consider to be a germ-line modification and therefore one that (in their view) leaves numerical identity untouched. They take an embryo that is a joint production of theirs, and use some genetic engineering technology to introduce a gene that confers deafness, before reimplanting the embryo in the womb. Here, the parents reduce the propensity of the individual to hear, and the resulting person—call her Kathy—might complain that her parents wronged her in so doing. After all, this looks like a case where her parents chose that Kathy be deaf. In the second case, a different couple uses an embryo selection technology, again because they want a child who is deaf. They generate a number of fertilized embryos, choosing to implant one only if it has genes conferring deafness. Bredenoord et al. seem to imply that in this case the resulting person—call her Kate—cannot complain that her parents did her any wrong with respect to her inability to hear, for Kate would not have existed had a different embryo been chosen for implantation. Kate's parents did not choose that Kate be deaf; they chose that Kate, rather than some different person, should come to exist. This is not an especially unusual thought: both McCarthy (2001) and Hope and McMillan (2003), among others, have assumed that embryo selection is identity-altering.

If the arguments of this chapter are correct, then origin essentialism supports no such contrast in our assessment of germ-line modification and embryo selection, at least not without considerable further argument. After all, while embryo selection affects a subset of the original materials of a person, the choice of which embryo to implant does not alter a vast array of non-embryonic origins of the resulting organism and neither does it order the broad timing of gestational processes.

We must be careful not to overstate the rather modest significance this point has for our evaluation of Parfit's (1984) 'Non-identity Problem', a problem with wide currency in discussions of reproductive ethics. The first thing to say is that Parfit himself, having recognized the problem, was unsure what to make of it. Parfit drew the conclusion that many of our actions do not wrong specific individuals, in the sense that we cannot truly say that a given individual's life would have gone better but for our action. His response to this was to look for an alternative conception of harm, as a way of making sense of the wrongs we do in the context of future generations. As he puts it, the area of morality 'concerned with beneficence and human well-being cannot be explained in person-affecting terms' (1984: 370–1). The upshot, for Parfit at least, was not to argue that we should somehow be less concerned if a course of action can be shown not to make any person's life go worse. He instead espouses something he calls the 'No Difference' view (1984: 367), which in turn suggests that we should not reassess our level of concern for some course of action, even if we can show that it is identity-altering. In his original presentation of the Non-identity Problem, for example, Parfit considers a policy he calls 'Depletion' (1984: 361–2). This involves the greedy consumption of natural resources in such a way that people's lives go very well in the short-term, but after that natural resources are so impoverished that future people have a poor quality of life. He argues that such a policy is identity-altering: it changes which people come to exist. One cannot say, then, that but for Depletion, the lives of those very future people would have gone better. Instead we should say that but for Depletion, those people would not have existed. And yet Parfit points out:

> We may have thought a policy like Depletion would be against the interests of future people. When we saw this was false, did we become less concerned about effects on future generations? [...] When I saw the problem, I did not become less concerned. (1984: 367)

Our first caveat, then, is to recall that on some views—including Parfit's own—the significance of the non-identity puzzle is not that it justifies some kind of contrasting ethical concern for those reproductive interventions that are identity-affecting, compared with those that are identity-preserving. Instead, the significance of Parfit's puzzle is that it shows the generic need within normative ethics for a defensible account of what Parfit calls 'Non-person-affecting Harm'. If we follow Parfit in embracing the No Difference view, we are unlikely to think that the question of which interventions turn out to be identity-altering makes much difference when it comes to an ethical assessment of some proposed set of reproductive choices.

The second caveat is to remind ourselves that we have not shown that all, or even many, reproductive interventions are identity-preserving. There surely are cases where our choices affect which numerical individual exists, rather than which qualitative traits some particular individual has. What is more, we can defeat parental essentialism and gamete essentialism simply by showing that an individual *could* have existed with different parents or that an individual *could* have existed with different gametes. This does not mean that an individual *would* have existed if his or her parents had acted differently in some respect or another, but it is this latter claim (or rather, its denial) that interests Parfit.

Parfit rests his presentation of the Non-identity Problem on what he calls the 'Time-Dependence Claim', and more specifically on the following relatively uncontroversial version of it: 'If any particular person had not been conceived within a month of the time when he was in fact conceived, he would in fact have never existed' (1984: 352). Parfit argues that the Time-Dependence Claim follows not only from origin essentialism, but also from other plausible views about identity across possible life histories. Parfit's famous case of the fourteen-year-old girl, for example, relies on the intuition that had this girl 'waited several years, she would have had a different child' (1984: 358). So much of the child's life and so much of its immediate history would have been different under these circumstances of delayed conception that no plausible account of identity—Kripkean or otherwise—allows that we could properly regard this as the same child.

The argument of this chapter does not trivialize Parfit's problem, but it does remind us that identity-affecting interventions may arise less frequently than many have thought. After all, when we choose which embryo to implant in the womb or which spermatozoon should fertilize an egg *in vitro*, we make a choice that does not alter the timing of conception in the manner that triggers Parfit's Time-Dependence Claim. The timing of conception would have been the same whichever package of genes had been chosen to contribute to the developing individual. Arguments about identity across possible worlds are slippery, and typically rest on questionable modal intuitions. Even so, for those whose questionable intuitions tell them that individuals' lives could have gone very differently, considerations of origin essentialism should not lead them to deny that these differences can include having come from alternative gametes and even from alternative parents.

PART II

Biology in Ethics and Political Philosophy

7
Development Aid
On Ontogeny and Ethics

7.1 Introduction

There is still a lot of talk about genetic traits, or innate traits, or the talents we are born with, or capacities and goods that are bestowed not by society but by nature.[1] In spite of all this talk, one thing is clear: regardless of the role genes might play in causing or even assuring height, or intelligence, or homosexuality, or stamina in running over long distances, babies are not literally born with these traits, at least not at the levels they attain later in life. The language of what we are born with is so common that one needs to be quite explicit in spelling out some trivial truths of ontogeny: newborn babies are never six feet tall; they cannot perform complex mathematical calculations; they do not harbour sexual desires directed specifically at humans of the same sex; they cannot run for several miles at speed without a break.

All of these capacities and talents must develop over some portion of the life of the person who possesses them. My grandfather would sometimes say of a talented football player: 'You can't learn passing like that, it's something you're born with.' The contrast is unfortunate. Learning is only one type of developmental process. One does not learn to be tall, yet

---

[1] This chapter was first published in *Studies in History and Philosophy of Biological and Biomedical Sciences* 33 (2002), 195–217. I received an enormous amount of help in writing it. A short version was read to the Department of History and Philosophy of Science, Cambridge University, in December 2001. I am grateful to the audience there, and also to Ron Amundson, André Ariew, Krister Bykvist, Soraya de Chadarevian, Anjan Chakravartty, Sohini Chowdhury, Marina Frasca-Spada, Anandi Hattiangadi, Adam Hedgecoe, Tor Lezemore, Serena Olsaretti, Martin Richards, and three anonymous referees from *Studies in History and Philosophy of Biological and Biomedical Sciences* for exceptionally generous and helpful comments on earlier drafts.

one's height is not something one is born with. Perhaps some football skills are similar: perhaps no one attains David Beckham's ability to cross the ball fast and low across the penalty area by practice alone. So the assertion that most talents and capacities must develop is compatible with the idea that they do not develop through a learning process, or that learning plays no important role in their development. Some traits we really are born with. Eye colour and limb number often go unchanged from birth to death. Yet even these traits must develop, although their development goes on inside the womb, rather than outside.

In this chapter, I consider what impact a proper recognition of the processes of development has on some traditional questions in ethics and bioethics. I begin in Section 7.2 by arguing against a form of 'genetic exceptionalism'. Genetic exceptionalists think that there is some special or distinctive feature of genes that justifies and explains our heightened concern for advances in genetic technologies and genetic knowledge.[2] I am especially concerned to argue against the view that genes are a special type of developmental cause. There is no ethically relevant principle that will distinguish the causal contribution to development of genes from the causal contribution to development of environmental factors. Although I think this holds true for all ethical problems including, for example, the question of whether genetic causes may be exculpatory, I focus here on questions of distributive justice. Arguments concerning these other problems must wait for another time.

In the third section, then, I argue that the 'parity principle' means that arguments in favour of the redistribution of environmental developmental resources—resources like schooling, nutrition, or the economic means thereto—should be expanded to include genetic developmental resources, and I reject a number of considerations that might be thought to exclude genes from the calculus of distributive justice.[3]

The upshot of Section 7.3's argument is that in appropriate circumstances genetic engineering may become one more tool for the distribution of developmental resources. Let me contain one objection

---

[2] See e.g. Murray (1997), Holm (1999), and Richards (2001) for related challenges to exceptionalism.

[3] Buchanan et al. (2000) offer similar views, although they arrive there via a different route, and they would reject some of the arguments I raise against genetic engineering in the second half of the paper.

at the outset. I will not be trying to argue that all genes are appropriate for distribution, just as I will not argue that all forms of nutrition should be distributed nor all forms of medical care. The question of which genes, or which foods, are appropriate for distribution turns, in part, on exactly what theory of justice we hold. A view that stresses the value of autonomy, say, might make available to all only a basic minimum set of resources to ensure development of the physical and mental capacities needed for the exercise of autonomy. Genes for intelligence (if such there be) and basic nutritional resources might be included in such a minimum set, but not genes for pole-vaulting nor foods like caviar. One who thinks we should strive to equalize welfare will more often allow that any genes, or any foods, that contribute to welfare are appropriate for distribution.

I should also point out that there are theories of justice that will not advocate the distribution of any genes, but that is because they do not advocate the distribution of any developmental resources at all. On some very narrow conceptions of equality of opportunity, for example, what matters is, roughly, that those whose talents best qualify them for jobs get those jobs. The further question of who has access to the resources that enable valuable talents to develop is of no ethical interest. Such a view tells us that development does not matter and that genes, along with nutrition and education, should not be distributed.

Henceforth I will be assuming that justice should be concerned with resources that enable adult traits to develop. Yet it is worth pointing out that those theories that say otherwise are congruent with the core of my argument that we will find no ethical distinction between genes and other developmental resources. Theories of justice should either think of all kinds of developmental resources as appropriate for distribution, or none.

In Section 7.4 I outline some good reasons for cautioning against over-enthusiastic genetic engineering. Some are practical—the genome may not be an effective place to intervene—yet others are deeper and remind us to be wary of the power of genetic engineering to yield superficial technological fixes to deep social problems. None of these arguments trades on exceptionalist principles—indeed, they show how the parity principle can help illuminate the concerns of groups such as disability rights activists.[4]

---

[4] For more detail on this, see Amundson (2005).

In Section 7.5 I return to the topic of exceptionalism, exploring a cautious version of the principle that I call 'exceptionalism by degrees'. The fact that genes do not pose problems of a different type from those posed by environmental factors leaves open the possibility that genes and genetic technologies could present familiar problems to degrees as yet unseen. I conclude by offering some brief comments, made from the perspective of the parity principle, to allay the fears of those who think that in advocating the distribution of genes I am backing a form of eugenics.

7.2 Genes, Development, and Genetic Exceptionalism

For many biologists and philosophers, the concept of innateness, or the related attempt to distinguish between nature and nurture, is viewed as a hopelessly confused territory that is best abandoned (e.g. Oyama 2000a, 2000b; Griffiths and Gray 1994).[5] I propose to begin, instead, by explaining what is meant by the more widely accepted concept of a genetic trait, specifically with the concept of a 'gene for' some trait. My intention here is not to undermine or provide a reductio of the 'gene for' concept. My goal is to give a deflationary account of that concept that preserves its role in genetics while exposing the similarities between genetic and environmental causation in the construction of organisms.

What does it mean to speak of a gene for some trait? What does it mean, for example, to say that someone has a gene for blue eyes? It must mean, at a minimum, that the gene plays some causal role in the development of the trait. Yet it cannot mean that the gene is the whole cause of the trait. No portion of DNA alone suffices to bring blue eyes into existence. Genes only cause traits to come into existence against a rich background of other developmental resources, just as pulling the trigger of a gun only causes death against a rich background of murderous resources—a loaded barrel, a keen aim—that are themselves efficacious causes only against the background of a pulled trigger. Our daily talk of causes does not require that causes bring about their effects alone,

---

[5] See Ariew (1996, 1999) for an attempt to construct a load-bearing concept of innateness out of Waddington's concept of canalization (Waddington 1957).

and a recognition of the causal influence of genes does not require that we deny the causal efficacy of non-genetic factors that are the background to this influence.

Accounts of causation are many and contentious, but a 'chance-raising' account, briefly and loosely sketched, will suffice for the elucidation of the causal role of genes in development.[6] Say that P causes Q just in case P raises the chances of Q against some set of background circumstances. That is, say P causes Q if and only if, in the circumstances, the chances of Q are greater with P than without. In the circumstances of a loaded gun, Morris's murderous intent, and a good aim, Morris's pulling the trigger raises the chances of death, hence it is a cause of death. Of course it is not *the* cause of death: loading the gun also causes death, as does taking aim. The death has many causes.

So far we have said what it is for some particular token gene—some individual strand of DNA—to have an effect. Yet the 'gene for' locution goes further than this. A gene for X picks out some collection of DNA strands, all of which are united by their molecular sequence. Does it follow that all of these strands have the same effect? No: as we have seen, the effects of genes can vary depending on the context they are in. The same is true of trigger pullings, considered as a type of cause. We can imagine circumstances where a trigger is pulled by a wonder marksman—call him Simon—whose calling is to save people from assassination by firing so as to intercept the bullets and arrows of others, thereby knocking them from their fatal paths. When Simon pulls the trigger, the effect is to preserve life, not to destroy it. When the actual background context varies, the actual effects of trigger pullings can vary also. The same is the case for genes. Many genes show the phenomenon of epistasis: their effects depend on what other genes are present. And, of course, gene effects can also depend on environmental context. So there is no guarantee that stretches of DNA of a certain type will have any common effect.

Murderous Morris's trigger pulling and Simon's trigger pulling do have similar proximate effects. They both cause a bullet to fly out of the barrel of the gun. The same may be the case with two token genes of the same sequential type. Even in quite different developmental contexts,

---

[6] See Sober (2001) for an excellent account of genetic causation, and Mellor (1995) for a thoroughly worked out account of causation as chance raising.

they might have the same effects on primary transcribed sequence, say. Yet their effects further downstream might be quite different. A gene for blue eyes is a type of DNA sequence whose tokens, in actual circumstances, usually raise the chances of having blue eyes. Talk of 'genes for' is thus doubly probabilistic. We learn that some stretch of DNA usually raises the chances of some phenotypic effect, although circumstances may be such that in some cases it lowers the chances of that effect and raises the chances of others. In other words, we learn a statistical fact about the typical effect, and we learn a further probabilistic fact that the stretch of DNA raises the chances of this outcome. My account here follows that of Sterelny and Kitcher, although space and purpose mean my discussion is less precise. As they say: 'the intuitive idea is simple: we can speak of genes for X if substitutions on a chromosome would lead, in the relevant environments, to a difference in the X-ishness of the phenotype' (Sterelny and Kitcher 1988: 162).

This concept has a certain amount of slack, to be taken up by pragmatic considerations. Sterelny and Kitcher index the 'gene for' concept to some environment. But which one? If any hypothetical environment is allowed, then a given sequence will be a gene for almost anything, depending on how exotic we allow our imagined genomic and extragenomic environments to be. It is in the interests of sensible research not to index gene effects to environments that are never instantiated or which are instantiated only in freakish circumstances.

I suspect instead that when many scientists talk of a gene for X-ishness, what they mean is that on average, in actual environments, the gene makes a causal difference to the X-ishness of the phenotype. Under these circumstances we notice that what a gene is for depends on which population, or sub-population, we are considering. Imagine some sequence type A, whose tokens make 2 per cent of the members of a population short and make 98 per cent of the same population tall. Perhaps the 2 per cent whom the sequence makes short have a very unusual diet, so that the sequence has a correspondingly unusual effect— just the opposite of its more normal effect. Now is this gene a gene for tallness? Across the entire population—short and tall people combined— yes, because its average effect is to increase tallness. Across the small population whom it makes short, no, because its average effect across these people is to make them short. If one picks a given token gene in one of the unusual 2 per cent and asks whether it is a gene for tallness or shortness, there is simply no answer unless one also specifies which population one is considering as the arbiter of typical effect (see also Sober 1988).

Let us imagine that the gene's effect is indeed dependent on diet. The unusual 2 per cent eat a food X so rich in some rare chemical that it blocks the synthesis of a growth hormone produced by A. What would make them tall is a variant gene B that would make the 98 per cent short, because the normal diet of the majority contains food Y, unknown to our nutritional minority, that interacts with B to block the synthesis of growth hormone. Now we can also imagine that the interaction of food X with gene A and of food Y with gene B is a purely deterministic affair. X and A or Y and B guarantee shortness. X and B or Y and A guarantee tallness.

Suppose that many of the nutritional majority acquire B and as a result become short. Their chances of being short, so long as they don't change their diet, are raised from zero to one by the mutation from A to B. Equally, many of the nutritional minority acquire B and as a result they become tall. Now a statistical study might show that across the entire population 90 per cent of those who have B are also short. It would clearly be wrong to assert in these circumstances that having B confers a 90 per cent chance on the bearer of being short. Having B never does this. B either confers a 100 per cent chance on the bearer of being short, or a 100 per cent chance of being tall. Talk of genetic 'risk factors' can obscure this important fact about causation. The typical effect in our population of a substitution from A to B is to cause shortness. Yet this does not mean that for each of the members of the population, substituting from A to B increases their chances of being short. We must remember that talk of risk factors across a population is often ambiguous between making a claim about quite uniform probabilistic effects of some gene and the statistically varying deterministic effects of that gene.

The sensitivity of genetic effects to context is depicted in genetics using the concept of the norm of reaction (Lewontin 1974). Norms of reaction are illustrated as graphs showing how organisms with different genotypes react in different environments. Such graphs can take all kinds of shapes. Consider some simple trait like height. The reaction graph may show some gene regularly associated with increased height across two different environments, as we see in Fig. 7.1. Here, G_1 always makes its bearers taller than G_2.

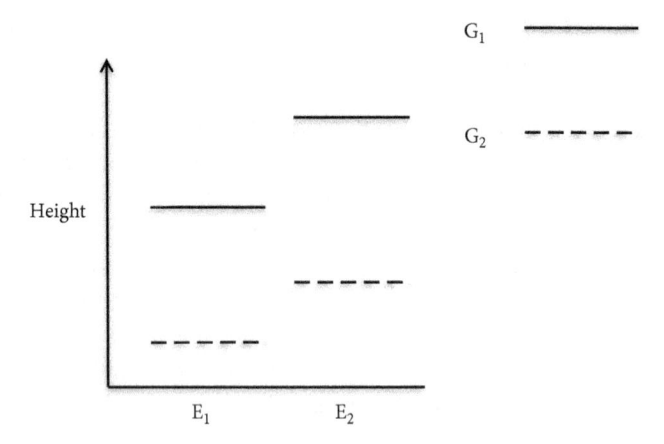

Fig. 7.1 Norm of reaction graph for height

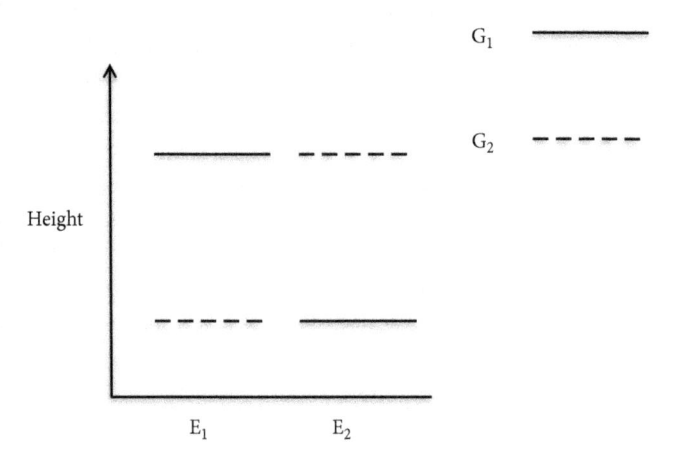

Fig. 7.2 Norm of reaction graph showing sensitivity of gene effect to environmental context

Alternatively, genotypes can show different norms of reaction, such as those illustrated in Fig. 7.2. Here, G_1 makes its bearers tall in one environment, but short in the other. According to our definition, we might say that in a population that includes only members in E_1, G_1 is a gene for tallness. Yet if we look at a population that includes only members in E_2, G_1 is a gene for shortness.

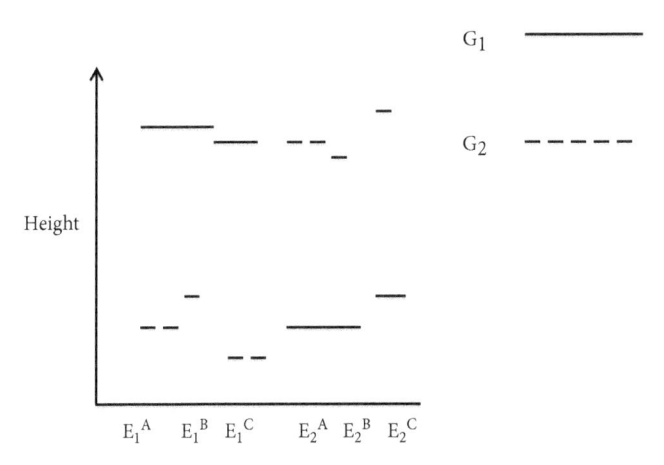

Fig. 7.3 Norm of reaction graph with subdivision of environmental context

Norms of reaction show the average value of some phenotypic characteristic, for phenotypes with some genotype, across a range of environments. It is a matter of theoretical significance how one decides to individuate environments. In Fig. 7.2, E_1 and E_2 could perhaps be further subdivided. So suppose our graph represents the height of plants and that E_1 represents some quite broad range of concentration of soil manganese, as does E_2. It could turn out that were we to divide E_1 and E_2 into smaller ranges, we could get a graph that vindicates our original choice of range, as shown in Fig. 7.3. Yet our findings could also show the initial partitioning of environments to be arbitrary. This possibility is illustrated in Fig. 7.4.

Fig. 7.4 shows that the average effect of G_1 on height in E_1 hides enormous variance within the parameters marked out by this choice of environmental range. Speaking of the average effect is liable to be rather misleading in this kind of context, for it may imply not only that a certain kind of effect is the mathematical average but also that it is widely found across the range in question. Yet Fig. 7.4 shows that the effect of G_1 on height varies considerably within E_1.

We must be sure, when evaluating claims that some gene is a gene for some trait, to clarify whether that claim refers to average effect, or typical effect, across some environment. A gene may have an average effect on height while having no typical effect whatsoever. Needless to say, a gene with some average effect on phenotype across the population might in

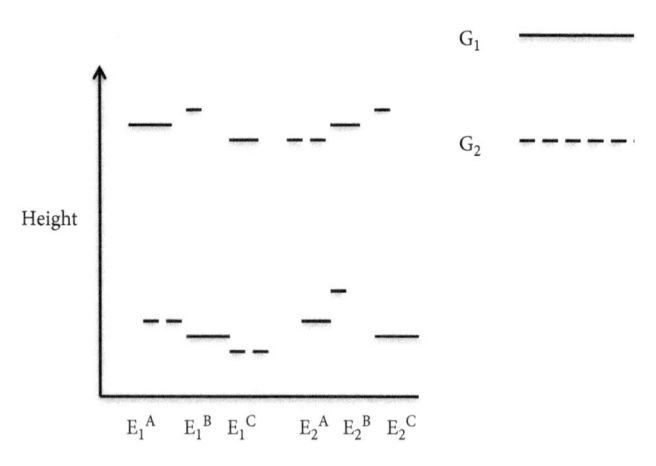

Fig. 7.4 Norm of reaction graph showing that average effect within E_1 and E_2 hides considerable sensitivity of gene effect to variation within those environments

fact be very unlikely to have that effect on any individual, and may even be unlikely to have an effect anything like its average effect.

Our discussion makes clear that, in the context of development at least, there is no general feature that distinguishes genetic causation and the developmental role of genes from environmental causation and the developmental role of environments. If the 'gene for' locution does nothing more than spell out the average, or typical, causal effect in some range of contexts of some gene, then there is an equivalent 'environment for' locution that spells out the typical causal effect in some range of contexts for some environmental factor. Smoking is an environment for cancer in just the same way that BRCA1 is a gene for cancer. Across a broad range of actual environments, smoking has both the typical effect of increasing the chances of an individual getting cancer and the average effect of increasing the chances of an individual getting cancer.

Perhaps there are genes for intelligence also. As I mentioned, we must be careful in interpreting the significance of any claim about 'genes for X-ishness' to understand whether evidence points to a typical effect of the gene in some broad range of actual environments or merely an average effect across environments that might be individuated in an arbitrary way. However the debate over the relation of genes to intelligence plays out (see Wasserman and Wachbroit 2001 for several good papers on behavioural genetics), we can be sure that there are environments for intelligence in just the same sense. Going to school, for example, or being read to have a typical positive effect on cognitive function across a broad range of actual contexts. Perhaps for some people they have no such effect, but, as we have seen, even claims about genes for traits at best pick out typical effects in a population.

It may seem that some genes have a particularly resilient effect on development. For some genes, their effects are the same regardless of how we try to alter developmental circumstances. This is the case for the Huntington's gene, for example. So far, we know of no environment in which the gene fails to cause the disease (excepting, of course, environments where no development occurs at all). Does this mean that we should acknowledge special developmental powers of genes? No. The Huntington's gene has some effect that cannot be eluded given our knowledge so far. Yet the same is the case with many environmental causes. Thalidomide, when administered to a growing embryo, has effects that cannot be overridden by later alterations to the developmental environment. And just as the Huntington's allele has an invariant effect in all known developmental environments that vary save for the presence of the allele, so thalidomide has an invariant effect in all known developmental environments—which in this case must be taken to include genetic environments—which vary save for the presence of the chemical. Environmental variables can have flat norms of reaction just as much as genes. There is no general way to distinguish the developmental actions of genes and the developmental actions of the environment.

Perhaps, then, some traits cannot be eluded, given our current knowledge. This is one widely used sense of the word 'innate'. Innate traits are those that show quite flat norms of reaction across known environments. Yet traits that are innate in this sense may not owe their variance in some population solely to the possession of genes. As we have seen, if the claim of the innateness of trait T is that all and only people with some gene G will develop T across all known developmental environments, then we can just as well say that some traits are innate in a parallel sense that all and only those people exposed to environmental factor X will develop T across all known developmental environments. What is salient in the claim of innateness here is not that variance is accounted for solely by the presence of some gene, but that variance is accounted for solely by the presence of some single, simple developmental resource across all known environments. The response that genes are not subject to human control will not do. The presence of many environmental agents, some of which may be unknown, is also often outside human control. At present we understand better how to modify environmental causes than genetic causes, yet no appeal to the determinants of innate traits will yield any sharper distinction between genetic and environmental causation.

Elements of the genome and elements of the developmental environment cannot be distinguished in terms of the kinds of effects they have on adult development. This conclusion has not required any controversial conception of development itself. A broadly interactionist view, where genes and environment are acknowledged to play mutually interacting roles, is the consensus position throughout biology, and it is all that we have required in order to state the causal roles of genes and environments in the construction of phenotypes.

Some recent theorists, most notably the developmental systems theorists (Gray 1992; Griffiths and Gray 1994; Oyama et al. 2001), have tried to go beyond the interactionist consensus to argue that genes and environments cannot be distinguished in their roles in evolution. This position is more controversial and requires a good deal of theoretical work in conceptualizing the processes of heredity, environment, and replication. Yet my conclusions for distributive justice rest only on the parity of genes and environments in development.

Just as no important distinction can be drawn between the causal contributions to development of genes and environments, so no important distinction can be drawn between the informational content of genes and environments—at least not in terms of the information these resources carry about phenotypic traits.

It is important to begin by distinguishing two senses of 'genetic information': first, we may simply refer to information about genes. In this sense, my genetic information is just information about my DNA sequence, and there is no more problem in speaking of it than we might have in speaking of information about the colour scheme in my house or what I had for breakfast. Yet questions of how such information should be stored, and who should be allowed access to it, are issues of considerable concern. There is a second sense of genetic information that attempts to elucidate the much stronger and more controversial idea that genes are a form of information. Some biologists and philosophers (e.g. Maynard Smith 2000) think that genes are a code of sorts, whereas few scientists (semiotic extremists aside) think that my breakfast, or the decorations in my house, are information or a code, even though there is plenty of information and code about all of these things.

It is uncontroversial, then, that there can be information about all kinds of developmental resources—genes, schools, money—but what of the view that genes themselves are a form of information? One often hears of genes coding for certain traits, of genes specifying certain traits, or of genes being blueprints for the adult phenotype. If this means simply that when holding environments constant phenotypes co-vary with alterations to certain genes, then we see immediately that environments are information bearers in the same sense. Against a constant genetic context, traits such as height co-vary with alterations to environmental factors such as nutritional regimes. So environments code for height in just the same way as genes (equally, they specify certain heights, or they are blueprints for adult phenotypes).

There are only two plausible attempts that I know of to flesh out a restrictive notion of information that would exclude the full range of developmental resources. Yet neither would give genetic developmental resources any privileged ethical status. Godfrey-Smith (2000a) has tried to sketch a concept of genes as code that restricts the coding role of genes to the primary post-transcriptional amino acid sequence they generate. On this understanding, neither genes nor environments code for phenotypic traits such as eye colour or intelligence, even though both may have a causal impact on such traits. Godfrey-Smith's view is, if anything, too restrictive and too grounded in the narrow project of uncovering the relationship between nucleotide sequence and amino acid chains to make genetic information of broader ethical significance.

The second main class of theories of genetic information looks beyond development to locate the informational content of genes in their evolutionary role. Sterelny et al. (1996), for example, argue that some developmental resources can be thought of as information bearers in virtue of their selective history. They argue that genes (and perhaps some other small subset of developmental resources, such as symbionts) have been selected for their role in guiding development, whereas standing environmental conditions, say, have not. This account has met with some resistance (Godfrey-Smith 2000b; Griffiths and Gray 1997). It is not clear, for example, how insisting that developmental information bearers are those resources that have been selected for their role in guiding development can exclude many of the environmental conditions that we have been considering, such as nutritional regimes. Food preferences can be selected for their developmental roles, as Sterelny et al. (1996: 66) acknowledge. But if food preferences can be selected, why not foods themselves? It seems appropriate to say that the diet of the calf has been selected for its developmental role, just as the shell of the hermit crab has been selected for its protective role.[7]

Perhaps there is a way for Sterelny et al. to exclude environmental developmental resources such as nutrition from the category of information bearers. In any case, it suffices here to point out that by making the status of genes as information bearers depend on their history, it becomes hard to see how the privileged status of genes could have ethical significance.

For Sterelny et al. it is neither the current structure of genes nor the nature of the effects that genes have on development that distinguishes them from environments. Suppose, then, that the entire human species, as it now is, had been created by chance from an unlikely coming together of inorganic matter. In this scenario, although both genes and environments would act in just the same way to produce adults as they do now, neither genes nor environments would carry information about phenotypes, for neither genes nor environments would have any history of selection.[8] This is not intended as an objection to their account. There may be good theoretical reasons within evolutionary biology for reserving the term 'information bearer' to denote an item with a certain kind of history. Yet the thought experiment of the instant human society does show that, whatever the successes of Sterelny et al.'s view, the status of some developmental resource as an information bearer in their sense cannot have any important ethical consequences.

The key result of this section is that from the perspective of development, even if not from the perspective of evolution, we have so far failed to find any significant difference between genes and environments. All have effects on development, of varying kinds, with varying regularity

---

[7] I am grateful to an anonymous reader of an earlier version of this chapter for suggesting this line of argument.

[8] Readers will recognize that this is the familiar 'Swampman' counterexample to teleosemantics (see e.g. Millikan 1984; Papineau 1993) turned against Sterelny et al.'s teleological conception of genetic information.

across contexts. Among both genetic and environmental factors, some have quite invariant effects across a broad range of actual contextual backgrounds, while others have effects that are quite sensitive to background. Both genetic and environmental factors should, I suggest, be conceived quite generically as 'developmental resources'—as items that make causal contributions of varying kinds to the development of the phenotype. Even drawing a distinction between genes and environment is somewhat artificial. 'Environmental factors' are defined only negatively as all those developmental factors that are not genetic. They therefore include many biological factors, even non-genetic resources that are carried in the gametes. One could just as well fix on a certain non-genetic type of developmental resource—nutrition, say—and define 'environmental factors' as all developmental resources that are non-nutritional. Here, genetic factors will be included as part of the environment of nutrition.

All of this makes a strong case against 'genetic exceptionalism': if genes and environments are no different in terms of their effects on development, it suggests in turn that the new genetics presents no new ethical problems that have not been encountered before, through our long-standing practices of modifying the developmental environment. Yet the implications of the parity thesis are not all deflationary. We tend to think that we have a duty to provide adequate developmental resources to all. If genes and environments do not differ in any relevant respects in their status as developmental resources, then many good arguments in favour of the redistribution of educational or nutritional resources, say, are also good arguments in favour of the redistribution of genetic developmental resources through genetic engineering. Of course we have not yet considered every attempt to draw an ethical distinction between genetic and environmental resources, nor could we; in the next section I consider some of the most plausible exceptionalist principles.

7.3 Some Objections Rebutted

The question of how distributive justice should proceed is, of course, contested. Should we aim for equality of resources, for equality of welfare, or for equality of some type of opportunity? Perhaps equality should not be our goal at all, but we should aim instead to ensure that all have some minimum of resources, welfare, or opportunity.[9] Most broadly egalitarian theories of justice will agree that we have some duty to distribute developmental resources such as nutrition and education. Nutrition and education are trivially resources, for example. Since nutrition and education are also uncontroversial contributors to both welfare and opportunity, it seems that these three theories of the currency of justice—welfare, resources, or opportunity—will agree that such developmental resources are the kinds of things that should be spread fairly around society.[10] But genes are developmental resources just like nutrition and schooling. So, since we think nutrition and schooling should be included in the calculus of distributive justice, we should include at least some genes in this calculus too. In this section I consider four alleged features of genes that might be thought to exempt them from considerations of distributive justice.

7.3.1 Genes cannot be altered, other developmental resources can. Only controllable resources can fall within the scope of distributive justice

Two comments suffice as a response here. First, if germ-line engineering technologies become available, genes will be subject to alteration, and such alterations will become another method of redistributing developmental resources to be considered alongside improvements in schooling or nutrition. Genetic engineering is at present a blunt tool, but it may not remain so.

Second, the fact that the allocation of genes is largely out of our control does not entail that this allocation should not be taken into account when further developmental resources that are under social control are distributed. Suppose we think that all should have equal opportunity, and suppose also that our genes contribute to opportunity by contributing to intelligence or even height. An understanding of the non-controllable distribution of these genes becomes important when we come to decide how to allocate resources such as schooling and nutrition that are

---

[9] For a variety of responses see Dworkin (2000), Sen (1992), Cohen (1989), and Rawls (1971).
[10] I acknowledged in Section 7.1 that some versions of equality of opportunity will not license the distribution of any developmental resources, including schooling and nutrition. The same is the case for libertarian theories of justice that see all redistribution as theft.

controllable. On at least some theories of justice, those with poor genetic complements will be entitled to compensating shares of these other resources.

7.3.2 Our genes dictate our identity, other developmental resources do not

One may think that distributive justice is a matter of giving certain resources to individuals. On this view, one might also think that the allocation of genes cannot fall under the aegis of distributive justice because genes determine the identity of individuals: by altering the genes one does not thereby give the same individual a different set of developmental resources; instead, one changes the individual—one brings a different person, of a different type, into existence.[11]

Such an argument might be motivated by Kripkean concerns about the essentiality of origins. We could not have had origins other than those we did in fact have, says Kripke (1980); hence we could not have had different genes, since what genes we have is a matter of where we came from.

A full examination of the view that one possesses one's genes essentially is beyond the scope of this chapter (but see Chapter 6). It is a shame that Kripke says so little about the essentiality of origins. He is explicit that he thinks that one could not have had two different parents. He does not say whether one could have had one different parent, hence whether half of one's genes might have been different. Nor does he make it clear whether he thinks origins are essential because genetic material is essential or whether genetic material is essential because origins are essential. The latter position is a possibility. Perhaps Kripke, working in the 1960s and 1970s, thought that the only way one could change the genes of one's offspring would be to change one of its gametes, hence also changing the origins of the offspring. What would Kripke say, then, about a case where my parents conceive, with the same sperm and egg that went to make me, but where the nuclear material is subsequently removed and new material is inserted? Here the origins of the embryo are preserved, yet its genetic material is not. Is this embryo me, with new genetic material, or is it someone who is not me, who happens to have identical

---

[11] Many bioethicists have followed Parfit (1984) in flagging problems about distributive justice, genes, and identity.

origins? Again, Kripke does not consider this kind of case, nor does he say enough about why he believes in the essentiality of origins for us to make much of a guess about what he would say.

It suffices for my argument to show that some of an individual's genes could have been different, and this is compatible with a view that says that one possesses some very high proportion of one's genes essentially. It is extremely implausible to assert that, had a point mutation occurred to any single gene in the embryo that gave rise to me, then I would not have existed (unless, of course, the mutation threatens the viability of the embryo). It is not so implausible to say that I would not have existed had all or many of those zygotic genes been altered.

A human embryo is formed (usually) from the union of two gametes, one maternal and one paternal. The resulting cell—the zygote—then divides repeatedly to form a foetus, then a person. Now perhaps some alterations to this process cannot be undertaken, or imagined, without the actual or imaginary bringing into existence of a different person as a result. Perhaps it is incoherent to imagine that I could have had two different parents, for example. Is it incoherent to imagine that I could have had one different parent? That case is far from clear. Suppose my mother had conceived at the same time as she in fact conceived me, but with Ronald Reagan rather than my actual father. Suppose, to make the case clearer still, that my actual father had no children. Is this a world where Ronald Reagan is my father or is it a world where my mother gave birth to someone other than me? If the former, then this is a world where many of my genes are different from how they are.

While someone may be able to mount an argument to show that I could not have had different parents, such an argument will not show that I could not have had at least some different genes. Suppose my actual parents had conceived me at the time they did and that they then chose to use some cutting-edge technological procedure to have some of my zygotic DNA removed and a different small portion of DNA spliced in. No good Kripkean argument will show that this is a world where I do not exist—there is no reason to think that I have every one of my genes necessarily, just as I do not have the other, non-genetic resources packaged in the gametes that gave rise to me necessarily. No argument from the necessity of origins can save genes from the calculus of distributive justice.

7.3.3 Genes are given by nature; other resources are under social control

This objection has been covered already. If the claim that genes are given by nature means only that the allocation of genes is not under our control, then the responses given to 7.3.1 suffice to rebut it. On the other hand, the claim may be instead that because our genetic allocation is bestowed by the hand of nature, then we should not try to alter it. Such complaints about not 'playing God' are vacuous, and this vacuity can be seen clearly from the perspective of genes as developmental resources. Not so long ago, other developmental resources were seen as part of the natural order—the allocation of wealth, education, social standing, and other privileges—yet many now think it appropriate to take control of the allocation of these resources. Many theories of distributive justice seek to overcome the arbitrariness of these various social and natural lotteries, and it is hard to see any principle that would justify our refusing to even out only one such arbitrary allotment—the allotment of genes.

7.3.4 Genes are inherited across generations; other developmental resources are not

Perhaps genes should not be considered as resources for redistribution because the effects of gene substitutions on an individual are inherited in the individual's offspring; but the effects of substitutions to other developmental resources are not inherited in this way.

Why would anyone think this an objection to genetic engineering? If some developmental resources are to the advantage of their possessors, then it might even be a good thing that those resources need not be actively distributed anew to the possessors' offspring. It seems clear that what motivates the original objection is, in fact, an epistemological worry. We cannot be sure that gene substitutions will have beneficial effects, and we risk seeing the magnification in future generations of our foolish interventions today.

In fact, the apparent distinction between the permanence of alterations to genetic material and the transience of alterations to other developmental resources is illusory. Knowledge—a non-genetic developmental resource—can also be passed from parents to children. Many other environmental features, although not passed on from one generation to the next, are nevertheless stable across many generations, and hence their alteration can be expected to have a lasting impact on those generations. Urban regeneration projects can have this character, for example, and the widespread suspicion of 'social engineering' seems to trade on just the same worries that some have in relation to genetic engineering. 'Social engineers' are also thought guilty of misunderstanding the future effects of their heavy-handed interventions; one generation's utopian housing scheme is a later one's crime blackspot or crack den. We need to be sure that we understand the effects of genes, but such considerations will not serve to distinguish genetic and other interventions.

7.4 Tempering Genetic Engineering

Having reviewed some bad reasons for limiting the scope of distributive justice to non-genetic resources, let me review three good reasons for at least reining in our enthusiasm for the redistribution of genes and genetic engineering.

7.4.1 Genes may be a poor place to intervene

The mere fact that genes form a subset of developmental resources does not tell us that the redistribution of genes is the most effective way to ensure the just provisioning of developmental resources. In many cases environmental interventions, or the provisioning of enriched environmental resources, may be more direct and more effective means to our social and political ends. The example given by Dawkins (1982: 12) of the 'gene for knitting' makes this clear. In modern Western societies, women tend to knit far more than men. What is more, this trait is under genetic control. Population studies would show that the possession of two X-chromosomes raises the chances of an individual learning to knit.

Some popular writers (such as Ridley 1999) have taken Dawkins' example to show that we must not confound genetic causation and genetic correlation. They have taken it to show that just because there is a correlation in Western societies between having two X-chromosomes and knitting, we should not therefore conclude that there is any causal relation between the possession of two X-chromosomes and knitting. Of course we need to be wary of inferring causal relations from correlations (see Sober 1993), yet this kind of response misses the force of the example. There really is a gene for knitting in Western societies, yet it exerts its influence over the phenotype via an extremely complex causal path. The additional X-chromosome causes primary sexual characteristics, which in turn cause infants to be treated in certain kinds of ways by other members of society, which treatment in turn causes girls to be more likely to learn to knit than boys.

One reader has suggested that this should be understood as a case of gene–environment correlation: girls happen to be born into one kind of social environment, boys into another. There is no gene for knitting. My response here is to say that girls indeed acquire one kind of environment, boys another, yet the X-chromosome is, in part, responsible for the acquisition of that environment. Because the possession of the chromosome makes a difference to what kind of environment one grows up in, the chromosome also makes a difference to one's chances of being in an environment where one will be taught to knit. Hence, given my earlier definition of 'gene for X-ishness', this is a gene for knitting still, in spite of the fact that gene–environment correlation also explains girls' higher chances of knitting. I noted that a consequence of that definition is that what a gene is for depends on what population we are looking at. This phenomenon is clearly demonstrated here: as another reader has pointed out, before the middle of the 18th century the gene for knitting was on the Y-chromosome. At this time, hand-knitting was done predominantly by men.

Suppose we were to decide for some strange reason that all members of a society should be just as likely to knit as others. Genetic engineering would be one way of doing this. By giving everyone the same sex chromosomes, we might even out the likelihood of individuals knitting across society. Yet it is obvious that the genetic engineering route would be an absurd one to take in the achievement of our social end. Altering patterns of socialization would be far more sensible.

The metaphor of the genome as a blueprint is surely partly to blame for inflating the prospects for genetic engineering, since it suggests very specific control over the phenotype by the genotype that may hold for only a few gene–trait pairs or none. We do not usually think of nutritional regimes as constituting part of the blueprint for a developed adult, yet nutritional regimes have causal influence over development in just the same sense as genes do. Suppose, again, that we want to equalize propensities to knit across some population. By changing patterns of instruction we can in fact achieve far more specific control over this attribute than by altering chromosomes. Giving everyone an additional X-chromosome will not only alter their propensity to knit; it will also alter their propensity to acquire breasts. Altering patterns of instruction by teaching knitting to all members of a society will have far more discrete effects.

The fear of the advent of 'designer babies' also trades on the assumption that genetic control will be far more precise than the control we already exert over development through education, nutrition, and so forth. Parents already have designer babies, in the sense that they influence their schooling, their cultural diet, and so forth. There is no guarantee that genetic engineering will give us any finer control over development than we already have.

In using an example of whole chromosome substitution, the relative clumsiness of developmental effects produced by genetic engineering is perhaps exaggerated. Only future genetics will tell us whether any shorter sequences that we more usually think of as genes have the predictable, discrete effects on phenotype that would make altering them a suitable means by which to attain our distributive ends. It may turn out, for example, that the only changes to phenotype that we can hope to use genetic engineering for are the elimination of some diseases. The scope of genetic engineering as a tool for distributive justice may thus be empirically limited to a modest and familiar role.

Finally, as I noted in Section 7.2, if 'gene for X-ishness' means only that a gene has an average effect on phenotype, then unless one has quite detailed information about the genetic and environmental causal background of an individual, one would refrain from inserting a new gene into that developing individual. This is because the effect of the token gene in that context could be nothing like the average effect of genes of that type.

In summary, genetic engineering may see little uptake not because of broad ethical principles of development, but because of matters of efficacy and economics. This is a general feature of positions that argue against exceptionalism. If genes and environments are on a par from the perspective of development, then while the modification of genes will tend to be admitted on ethical grounds as no different from environmental modification, it will also tend to be played down on practical grounds as likely to be no more efficacious than environmental modification.

7.4.2 Genetic engineering may be used to bias in favour of certain functional modes

Let us imagine two genotypes, each of which contributes to intelligence with norms of reaction illustrated in Fig. 7.5.

Suppose we decide that equality of opportunity demands that all growing persons have equal developmental resources that contribute to intelligence. If a person is to be born into E_1 with G_2, say, it might seem a neutral decision whether to give that individual G_1, or to ensure that the individual is raised in E_2. It is possible, however, that although E_2 and G_2, and E_1 and G_1, yield identical levels of function, they may be underpinned by quite different modes of function.[12] The interaction of G_2 and E_2 might yield the same level of intelligence, but through a quite different set of mechanisms. Perhaps G_2 is a gene that results in disease in E_1. In E_2, on the other hand, some set of compensating resources and alternative developmental pathways becomes available, so that, while the individual's appearance, methods, and behaviours are quite different from those of members of the population with G_1 and E_1, the level of cognitive performance is not affected.

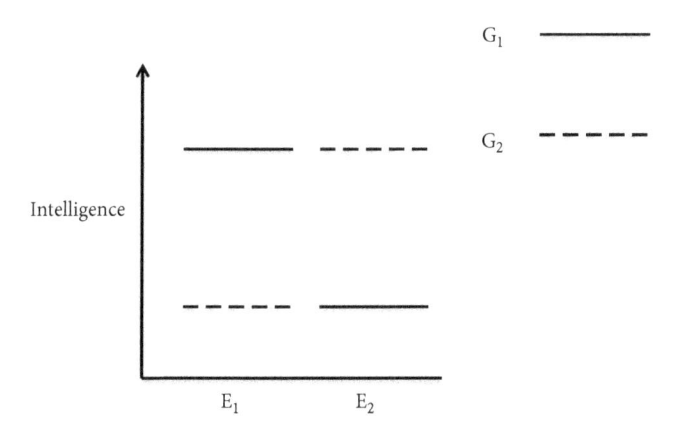

Fig. 7.5 Norm of reaction graph illustrating hypothetical contributions of genes to intelligence

---

[12] I owe the helpful distinction between functional level and functional mode to Amundson (2000). My discussion of genetic engineering in this section is heavily influenced by Amundson's work, and I am also grateful to him for advice on this section.

In some cases, the different environment might not be imposed from outside, but might instead be produced by the individual in question so as to compensate for the more normally encountered environment E_1. Amundson (2000) gives the example of autistics' behaviours of 'stimming'—rocking back and forth while sitting, tapping one's face, flipping fingers in front of one's face—that seem to enhance the functionality of the individual. He summarizes: 'A non-stimming autistic person may be more cosmetically normal, but able to function only at a lower level' (Amundson 2000: 50).

Why should these facts make any kind of case against genetic engineering? The answer is that when we have a choice over how to augment functional performance—either by genetic engineering or by environmental alteration—we should not be tempted to increase the most commonly found functional mode by insisting on the use of genetic engineering techniques. We should be especially wary in cases like this one of favouring genetic engineering over social or environmental alteration on the spurious grounds that more typical modes of functioning are healthier than atypical modes.

Even so, one may respond that this case only shows that genetic engineering and environmental alteration may be equally good distributive tools. It does not show genetic engineering to be wrong. By hypothesis, both functional modes yield equal intelligence and as a result equal opportunities. Hence the choice of the typical functional mode through genetic engineering over the unusual functional mode through environmental modification is ethically neutral.

Let me illustrate my response with an example. Consider a racist society in which black children tend to have far lower opportunity than whites owing to various forms of discrimination and abuse. The opportunity of a future child of a black couple can be increased either by altering the social environment so that whites and blacks have equal opportunity, or by altering the genes of the child so that it will grow up to be white. Here there is a choice for how opportunity is to be equalized, yet I suspect most of us have an intuition that this choice is not ethically neutral. And again, we would not condemn only genetic engineering—we would think a dietary or hormonal intervention to change skin colour would be equally wrong. Our intuition is that altering society so that diverse kinds of people can flourish is a better way to promote equal opportunity than altering individuals so that only one kind of person exists.

What this example helps us to see is that far from resulting in general advocacy of genetic engineering, the perspective of the parity principle helps us to articulate the suspicions of genetic engineering among groups such as disability rights activists. Their concerns turn on the fact that enthusiasm for genetic engineering over social and environmental alterations, such as the provision of wheelchair ramps for paraplegics, can be expressive of contempt for the value of diverse functional modes that rely on social reform for their full expression. Always to look to genetic engineering (or to a range of other medical interventions) to equalize opportunity, or resources, or welfare is to ignore social deficiencies that sometimes stand in need of remedy.

7.4.3 Genetic engineering may violate bodily integrity

Genes differ from developmental resources such as food and education in that they are internal to the body and their alteration requires invasive procedures. We might therefore think that this is a strong reason against using genetic engineering as one of our tools for the distribution of developmental resources, since any prima facie reason in favour of distribution through genetic engineering is countered by its violation of bodily integrity, either of the body of the growing embryo's mother or of the embryo itself. I can only counter this by saying, first, that many public health measures currently administered to infants—inoculations, for example—do demand invasive procedures and, second, that the question of whether the rights of the mother are violated must be balanced against the interests of the child to come. A thorough examination of these issues would take the argument too far from the main topics of the paper.

7.5 Exceptionalism by Degree

So far I have argued that the similar status of genes and environment as developmental resources demands that we include genes in our calculations for distributive justice. Perhaps it is true that genes and environments cause traits in just the same ways, and that there is no good way to distinguish the senses in which they carry information about the phenotype. It is a consequence of the parity thesis that the new genetics poses no new types of ethical problems. This said, genetic technologies enable such easy access to personal information that special concern is justified nonetheless. This is a position we might term 'exceptionalism by degree'.[13]

We have seen that both genes and environments contain information about phenotypes in the limited sense that both genes and environments can be used to predict the development of phenotypes against some set of background conditions. This means that we should be no more concerned about the storage of and access to some particular piece of genetic information than about storage of and access to some particular piece of environmental information. There is no important difference between letting an insurer know that one has a gene that disposes one to cancer and letting an insurer know that one was brought up in Cornwall, where a granite-rich environment also disposes one to cancer through radiation.

There are, however, reasons to be especially cautious of the information about us that is contained in our genome, primarily because of its ease of capture from a cell and because of the availability of cells. There is no environmental analogue to the cheek swab followed by complete sequencing of an individual's genome: there is no simple set of actions that can capture the totality of information regarding one's environmental developmental resources. Perhaps school records will give an indication of what one was taught when growing up, and a parent's supermarket receipts might give an indication of what the growing child ate. Yet one's environment simply cannot be sequenced in the same way, although information can be gathered about it.

Nor is one's environmental information as easily available as genetic information. Some of us may leave our supermarket receipts lying around, so that those who wish to pry can gain some idea of how many cigarettes we might be smoking, and hence what our chances of cancer are. Yet information regarding the totality of our environmental developmental resources does not drop off us and lie for all to see in the same way that information regarding our genetic developmental resources does every time we brush our hair. The problems that could arise as a result of the ubiquity and richness of genetic information are the subject of the science fiction film *Gattaca*. Here, a future society is

---

[13] Richards (2001) argues against genetic exceptionalism along much the same lines as I do. In advocating a mild form of exceptionalism, I do not so much argue against Richards as give more stress to the riders he puts on exceptionalism.

imagined where discrimination occurs on the basis of one's genome and where cells are regularly pilfered and examined to estimate a range of phenotypic traits of interest. It is hard to imagine *Gattaca* being made about discarded supermarket receipts, rather than discarded cells. Yet the difference between accessing genetic information and environmental information is one of degree, not one of kind.

It should be stressed, however, that even exceptionalism by degree is bought in part using the flimsy currency of the distinction between genes and environment. As we have seen, this distinction is in some ways arbitrary. Since 'environment' just means 'all developmental resources with the exception of genes', it is hardly surprising that no simple recovery procedure can draw together information on such a motley crew of resources. We should not compare genes against all other developmental resources to assess their exceptional status; rather, we should compare genes against other particular types of developmental resource.

Many of our concerns about increasing knowledge of genes and gene action derive from the possible predictive power that such knowledge might have regarding disease, intelligence, or criminality. It is no surprise to learn, then, that concerns about the storage of information about genes exemplify our more general concerns about the control and privacy of other kinds of predictive information stored in electronic format. There are databases that contain information at least as rich, and perhaps as open to access by inappropriate audiences, as any store of genes or information about genes. An insurance company in possession of an individual's bank statements, or perhaps even their internet 'cookies', could tell far more about the makeup of that person than if they possessed that person's DNA sequence—at least given our current knowledge of functional genomics. An itemized statement that describes exactly what kinds of products are bought in what shops, for example, is of considerable health interest. It would be quite possible to draw up sophisticated actuarial tables estimating risk for individuals who buy gym subscriptions, who make frequent visits to the pharmacist, who regularly travel abroad, who spend in greasy spoons and never in health food shops, and so forth. Perhaps it is hard to imagine a version of *Gattaca* that deals with the possible social consequences of discarded supermarket receipts; it is not so hard to imagine a science fiction film similar to *Gattaca* that deals with the social consequences that might derive from the ability to hack into the electronic records of any individual's spending profile over a period of years. And, indeed, Anjan Chakravartty informs me that there is such a film—*Conspiracy Theory*.

7.6 Conclusions: On Engineering and Eugenics

In this chapter I have tried to sketch some consequences for political philosophy and bioethics that derive from a quite uncontroversial interactionist conception of organic development. I have argued that from the perspective of development there is no good reason to distinguish the causal contributions of genes and environments. This has both inflationary and deflationary consequences. The recognition that genes are, like environmental features, a type of developmental resource constitutes a strong prima facie argument for their inclusion in considerations of distributive justice, yet it also reminds us that alterations to non-genetic resources may prove the most appropriate practical means to our distributive ends.

In advocating bringing the allocation of genes into the calculus of distributive justice, am I guilty of backing a form of eugenics? The question of what was wrong with eugenics is complex, and I will not say much about it here. I have already used the parity principle to caution against over-enthusiastic genetic engineering on grounds that have been brought against eugenicists also—namely, that engineering enthusiasts may be guilty of advocating superficial technological fixes to deep social problems.

The parity principle exposes the incoherent tendency of many policy advisory bodies to raise the spectre of eugenics when, but only when, new genetic technologies are under discussion. We are rightly concerned that outside agents—perhaps especially governments—should not be allowed too much control over how lives are fashioned. It is not only state-directed programmes for the alteration of the gene pool that should ring alarm bells. 'Social engineering' is also a dirty word in policy circles, and another element of our fear for both social and genetic engineering is that governments might gain too much control over what types of people exist by taking too much control over how developmental resources—be they genetic, nutritional, or educational—are allocated.

Genes should fall into the calculus of distributive justice, just as nutrition and schooling do. Yet there are limits to the control that governments can be allowed over the allocation of all of these resources. The parity principle tells us that the relationship between genetics and justice is no less complex than the relationship between justice and developmental resources full stop.

8

Prospects for Evolutionary Policy

8.1 Evolutionary Policy

Two well-known and widely accepted arguments seem to show that evolutionary psychology is unlikely to tell us much of relevance to public policy.[1] First, there is no good inference from the claim that X is an adaptation to the claim that X cannot be altered. So evolutionary psychology, even if it can establish that male philandering is adaptive, will not thereby tell us that male philandering is a trait that policy-makers must work around rather than seek in vain to stamp out (Kitcher 1985). Second, the heuristic value of attempting to shed light on the makeup of the mind by reflection on the demands of ancestral environments is likely to be quite weak. Traits cannot be predicted from rough-grained facts about ancestral environments alone. So evolutionary psychology is also unlikely to give policy-makers much of a leg-up in understanding how our minds work, even if it can explain why we have the kinds of minds we do (Sterelny and Griffiths 1999).

In spite of all this, a small number of biologists and philosophers continue to link the results of evolutionary psychology to matters of political concern. A few years ago, the UK policy think-tank Demos devoted its quarterly review to a series of short articles by journalists,

---

[1] This chapter was first published in *Philosophy* 78 (2003), pp. 496–514. Earlier versions were presented to the Cambridge University Moral Sciences Club in May 2000, and to the Cambridge University Faculty of Zoology in June 2001. I am grateful to the audiences there and also to Patrick Bateson, Oliver Curry, Dylan Evans, Paul Griffiths, Kevin Laland, Richard Lewontin, Sadiah Qureshi, Martin Richards, and Sharmila Sohoni for helpful comments and discussion. I am also grateful to Clare College, Cambridge, for support while I was writing and researching this paper.

philosophers, and scientists on the input that evolutionary psychology might make to policy and economics (Curry et al. 1996). Peter Singer (1999) has urged a move towards a 'Darwinian left' that refashions itself in response to our increasing understanding of human nature. Helena Cronin has written a number of pieces in collaboration with Oliver Curry (Cronin and Curry 2000a; 2000b), arguing that the UK government's policy on the family would benefit from incorporating evolutionary psychologists Daly and Wilson's work on homicide (Daly and Wilson 1988; 1998). Thornhill and Palmer (2000) believe that an evolutionary understanding of rape will overturn, or at least greatly enrich, conventional wisdom on how to reduce the incidence of rape. Occupational psychology has jumped on board, with the publication of a book arguing that managers must understand the ape within if they are to get the most out of their staff (Nicholson 2000).

Might evolutionary psychology have relevance to matters political after all? I suggest in this chapter that the two old arguments still suffice to rebut even these more recent attempts to link evolution and policy. In Section 8.2 I briefly consider (and reject) the claim that even though showing some trait to be an adaptation does not entail that it cannot be altered, it nevertheless constitutes strong evidence that the trait will be very difficult to alter through social and cultural means. Section 8.3 considers Singer's 'Darwinian left' and shows this position also to fall foul of some misunderstandings of adaptation. Section 8.4 runs quickly through a series of fairly well-known arguments that show why reflection on ancestral demands on organisms is unlikely to constitute any very strong heuristic for predicting what traits will have evolved in response to these demands. Sections 8.5 and 8.6 look at two case studies—one from Daly and Wilson's work on homicide, another from Thornhill and Palmer's work on rape—that seem to show that, philosophical arguments notwithstanding, evolutionary psychology has been of value in illuminating features of human thought and behaviour that would otherwise have gone unnoticed. While I do not try to argue against these authors' evolutionary explanations for why we behave as we do, I do suggest that evolutionary thinking has played only a minor role in establishing the existence of the behaviours in question and the mechanisms that underlie them. The cases also point to further worries in deriving policy recommendations from evolutionary psychology.

8.2 Limiting Ambition

Biologists, psychologists, and philosophers of biology almost never assert that adaptations cannot be altered. They certainly do not do so in the technical literature. The only explicit statement I have found of the view that human nature is fixed is in a popular piece by Cronin and Curry (2000a), who ask:

> if human nature is the result of evolution, aren't we stuck with it? Surely there is little that policy makers can achieve? Not at all. Human nature is fixed; but the behaviour that it generates is richly varied, the result of our evolved minds reacting to different circumstances.

Even here the proper interpretation of Cronin and Curry's statement is not straightforward. It is possible that in saying that human nature is 'fixed' they use this word in the more technical sense of population genetics to mean 'is represented at a frequency close to 100 per cent in the population'. The context of contrast with the possibility of flexibility in behaviour goes against this interpretation, so I will assume that by 'fixed' they mean something like 'inevitable'. Cronin and Curry agree that the behaviours generated by human adaptations can vary according to environmental cues; however, they assert that the underlying mechanisms that respond to these cues are fixed in this sense.

It is quite easy to see why this claim about the fixity of adaptations is fallacious. The claim (for example) that male promiscuity is an adaptation tells us that males who were promiscuous out-reproduced males who were not. It also tells us that there has been some strong resemblance in terms of promiscuity between males and their male offspring. Finally, it tells us that this correlation between parents and offspring in terms of promiscuity was reliable across the majority of environments in which males have, in their evolutionary past, developed. This claim does not entail that there are no novel circumstances, or circumstances that have been rare but could become common, in which males do not develop to be promiscuous, or in which women develop to be more promiscuous than men. A trait can be an adaptation, hence part of 'human nature', and also be subject to alteration by environmental manipulation. Of course Cronin and Curry may also believe that there are circumstances under which males will be less promiscuous than women; however, they seem to assert that the underlying evolved mechanisms in males and females that produce varied sexual behaviours in response to environmental cues cannot themselves be altered. Hence, they also appear committed to the view that policy needs to work in accordance with this nature that is 'fixed'.

One might respond to this rebuttal with an inductive argument: adaptations must have been developmentally resilient to commonly encountered perturbations in past environments. It is therefore highly unlikely that they will be alterable by perturbations in current and future developmental environments. This argument is also weak. It is true that we should expect adaptations to be buffered against commonly encountered variation in the environments in which they emerged. That is to say, we should expect adaptations to be well 'canalized', for otherwise a beneficial trait would be almost certain to fail to develop because of its sensitivity to the inescapable perturbations of developmental environments (Waddington 1957). Yet natural selection cannot build traits which are impervious to novel developmental circumstances. Since we have the power to design novel educational, cultural, and even nutritional developmental environments we might succeed in disrupting the normal adaptive outcome and producing the traits we desire.

Most evolutionary psychologists would accept, and have defended, arguments of the sort laid out in the previous paragraph. Tooby and Cosmides, for example, see clearly that the evolutionary view of the mind leaves open the possibility of the alteration of adaptive traits:

Developmental processes have been selected to defend themselves against the ordinary kinds of environmental and genetic variability that were characteristic of the environment of evolutionary adaptedness, although not, of course, against evolutionarily novel or unusual manipulations. (Tooby and Cosmides 1992: 81)

The last two words of this quotation are relevant to the policy agenda. Just as selection cannot build traits to be impervious to novel environments, it need not build traits impervious to developmental perturbations that exist rarely in the environments in which the trait evolved. This means that we need not look only to very new environmental modifications if we wish to avoid traits that we think are adaptations. Modifications that have been rare in the past might also do the job. What for the evolutionist is merely developmental noise that results in a small fraction of women developing traits like excellence in spatial reasoning, desires for multiple sexual partners, lack of interest in raising a family, or the drive to dominate and acquire power in a competitive workplace might, for some feminists, be the developmental key to empowering women. There is, then, nothing wrong even on evolutionary grounds with feminists, say, looking to the educational or cultural environments of those women whom the evolutionist might view as wholly atypical of their sex for clues as to how more women in the future might develop traits which these feminists view as admirable.[2]

So far I have been assuming that selection will ensure the reliable development of adaptations by creating developmental pathways that are insensitive to variations in local environments. It is worth noting that theoretically, at least, this is not the only way that reliable development can be assured. Another possibility lies in the organism minimizing the kinds of perturbations that can occur in local environments by constructing and maintaining those environments itself. This phenomenon of niche construction is well documented throughout nature, and it offers organisms the possibility of assuring inheritance of traits across generations by engineering developmental environments that are conducive to the development of those traits (Odling-Smee et al. 1996; Laland et al. 2000). Sterelny (2001) gives a number of examples of niche construction outside of human society. To take just one illustration, beavers make lodges, and these lodges provide shelter against extremes of climate and protect against predators, thereby facilitating the development of future generations of beavers.

Humans are masters of niche construction. Our buildings, customs, and libraries can all be thought of as environmental features that are built and maintained by us, yet which also ensure the continued development of future generations able to sustain these very features. On a view of inheritance that stresses the role of niche construction, the generation of social institutions becomes a method for ensuring that developmental environments are constant, and hence that various adaptive traits will continue to be produced. In other words, if we take human niche construction seriously we also have to take seriously the view that

---

[2] Radcliffe-Richards' position on this question is not entirely clear. She notes that unfamiliar forms of culture or education might alter adaptive traits. However, she also suggests that 'Gould is at least close to the truth in saying that if sociobiologists are right in saying that certain characteristics are of an evolutionarily useful type, we shall not be able to alter them "either by will, education or culture"' (Radcliffe-Richards 2000: 117).

the stability of various social and cultural institutions is an important means by which adaptive traits are preserved. If we grant a fairly strong role to niche construction, we are returned, on biological grounds, to a traditional view that explains the stability of many human traits by reference to culture and society, and that leaves open the thought that revolutionary changes to culture and society can disrupt this stability and redirect human development.[3]

8.3 The Darwinian Left

In his short book *A Darwinian Left*, Peter Singer (1999) suggests that the left needs to revise some of its views to take into account the results of evolutionary psychology. Singer understands that adaptation does not entail fixity. Even so, I'm not sure that Singer takes this lesson to heart. In this section I try to show how the left emerges more or less unaltered by the results of evolutionary psychology.

According to Singer a Darwinian left would not, among other things:

- deny the existence of a human nature, nor insist that human nature is inherently good, nor that it is infinitely malleable;
- expect to end all conflict and strife between human beings, whether by political revolution, social change, or better education;
- assume that all inequalities are due to discrimination, prejudice, oppression or social conditioning. Some will be, but this cannot be assumed in every case. (Singer 1999: 60–1)

The first of the three claims is extremely vague. At the beginning of his book, Singer offers the following quotation from Marx, as an example of what he is arguing against: 'the human essence is no abstraction inherent in each single individual. In its reality it is the ensemble of the social relations.'

---

[3] Segerstråle (2000: 92) cites a terrifying example of ignoring the gap between demonstrating adaptation and demonstrating resistance to alteration due, apparently, to Senator J. William Fulbright at the 'Man and Beast' conference in Washington, DC in 1969: 'If we assume that men generally are inherently aggressive in their tendency [...] if this is inherent and man cannot be educated away from it, it certainly makes a great deal of difference in one's attitudes toward current problems [...] If we are inherently committed by nature to this aggressive tendency to fight, well then, I certainly would not be bothering about all this business of arms limitations or talks with the Russians.'

The claim that human nature is socially constructed, or that it is the 'ensemble of the social relations', is not equivalent to the claim that human nature does not exist. Yet it does suggest that by altering social relations, human nature may be altered. That is a position perfectly in tune with the evolutionary view of human nature as a series of malleable adaptations.

Singer's second claim either amounts to the truism that ending conflict and strife will be difficult or it commits the fallacy of thinking that if violent and aggressive tendencies are evolved, then they cannot be removed by political revolution, social change, or better education. It is important to note here that even views of our nature as in some sense socially constructed, and hence which are typically thought to be opposed to evolutionary psychological views, also leave open the possibility that some traits may be either hard or impossible to alter, given our current state of knowledge. Suppose we think that the failure of women to attain senior positions in large corporations is a result of socially conditioned prejudices among the men in charge of making appointments. This view does not entail that these prejudices are easy to alter, especially given our scant knowledge of the means by which patterns of social conditioning might be changed. Cultural practices can become entrenched in systems that are of purely social origin, and this entrenchment can make them just as hard to shift as any biological adaptation.

Singer's third claim is also ambiguous. It is unclear what the remaining possibility for the existence of inequalities is supposed to be. Human nature? Innate abilities? The claim that a trait is an adaptation, and in this sense part of human nature, is compatible with the claim that the trait owes its reliable development to reliably recurring forms of social conditioning. Hence, inequalities that are, in this sense, the result of human nature might also be the result of social conditioning.

One important political point that does emerge from discussion of Singer is that if we are interested in modifying or altering those traits that we find undesirable, then it may turn out that this cannot always be done through educational programmes. We need to know how traits develop if we want to alter them, and some traits may not be sensitive to alterations in educational or social circumstances, but may instead demand pharmaceutical, nutritional, or other interventions. Perhaps this is what Singer means in his third comment above. If so, then I am wrong to attack him. Yet if this is what Singer means, we should note that the faith of the left in the ability of education to modify all traits will not be eroded primarily by evolutionary work, but rather by work in developmental psychology.

As a final comment on Singer's third claim, we should recognize (as Singer surely does, too) that even if some aspects of human nature turn out not to be caused by social conditioning, it will often be the case that the ills these aspects of human nature give rise to are in part the result of conditioning. Perhaps the fact that more chief executives are male than female is a result of the fact that women have an evolved tendency to want to care for their children that cannot be altered by educational manipulations. Still, one can say that the disparity also owes itself to social conditioning that tells us that crèche facilities should not be made available in the workplace or that workers should not bring their children to work.

In brief, the contrast between a Darwinian and a non-Darwinian left is elusive. Since evolutionary explanations are typically silent about development, evolutionary views of human nature are compatible with views that most of the ills in human society are results of social conditioning. Of course this latter view may be wrong, but what will tell us that it is wrong for certain cases will be research into sociology, developmental psychology, and developmental genetics, not evolutionary psychology itself.

8.4 Finding Patterns and Mechanisms

Some think that evolutionary psychology is essential for understanding our cognitive dispositions, the patterns of behaviour that those dispositions yield, and the ways to control the expression of those dispositions. Cosmides et al. (1992: 9) outline one set of reasons for having such hopes: 'By understanding the selection processes that our hominid ancestors faced—by understanding what kind of adaptive problems they had to solve—one should be able to gain some insight into the design of the information-processing mechanisms that evolved to solve these problems.'

It would be foolish to argue that this kind of thinking—usually called 'adaptive thinking'—never has borne fruit and never could. To argue for the first claim would demand an exhaustive survey of the evolutionary psychological literature, and to argue for the second would be rash.

The evolutionary perspective encourages the generation of functional hypotheses where we would otherwise be inclined to ignore the possibility that a trait may provide some benefit to the organism. Darwinian medicine provides a good example here, where the investigator considers whether traits usually considered diseases, such as mild depression, might not be adaptive. Conversely, the evolutionary perspective also encourages us to consider hypotheses of conflict—between the sexes, between parents and offspring, between siblings—where other perspectives may too easily assume that the relationships must somehow be harmonious. Perhaps sometimes reflection on past selection pressures (adaptive problems) will indeed lead us to formulate hypotheses about cognitive architecture that we would have been unlikely to have considered otherwise. And sometimes these hypotheses will be confirmed. What is more, it is clear that the kind of research that one needs to do to confirm the existence of such putative adaptations can sometimes yield results that are relevant to policy decisions. That will be so even when proposed adaptations are not observed. Daly and Wilson's work on the evolution of violence requires that they undertake research on patterns of violent activity in modern societies. And this research could, of course, lead to the identification of significant patterns even when their evolutionary hypotheses are put in doubt.

In spite of these caveats, adaptive thinking is likely to remain a fairly weak predictive tool for uncovering the workings of our minds. That is because structures likely to evolve in a population cannot be read off from rough-grained features of that population's environment (Lewens 2002; Sterelny and Griffiths 1999; Davies 1999). We also need to know what range of traits is available for selection; mice will not evolve micromouse machine guns to deal with the problem of predation by owls. We need to know how traits affect fitness in some detail; a mouse with ears so big that owls can be heard at a great distance will suffer if those large ears also make the mouse conspicuous in the grass. Finally, we need to know how traits interact with each other; if mouse development is set up so that big ears are always accompanied by tiny legs, then the advantages conferred by hearing will be outweighed by the disadvantages of slow running. Needless to say, we will rarely have access to this information for ancestral human populations. If we do—if, for example, we know how cognitive traits interacted with each other during development, what kinds of constraints operated on the development of traits, and how those traits affected fitness—then it is likely to be because we have already undertaken broad investigation and experimentation with related populations, perhaps even with current human populations. A fairly deep understanding of the development and workings of the ancestral human mind is a prerequisite to generating accurate predictions about what the long-term effects of evolution in ancestral environments might be.

This kind of objection tends to infuriate evolutionary psychologists. Cosmides and Tooby (1997) remind us that our knowledge of the way of life of our ancestors is really quite rich:

> Our ancestors nursed, had two sexes, hunted, gathered, chose mates, used tools, had colour vision, bled when wounded, were predated upon, were subject to viral infections, were incapacitated from injuries, had deleterious recessives and so were subject to inbreeding depression if they mated with siblings, fought with each other, lived in a biotic environment with felids, snakes and plant toxins, etc.

All these things are true, but still a difficulty remains in deciding how to translate a description of a physical environment into a set of predictive adaptive problems. To take the case of plant toxins, should our ancestors have evolved a general horror for novel plant species, should they have evolved means of removing the toxins from plants by cooking them, or should they have evolved specific and quite different horrors of varying intensities for different types of toxic plant? If we are to give our catalogue of adaptive problems strong predictive power, then we need to know whether the problem faced by our ancestors was that of eliminating toxins from plants, of steering clear of nightshades, or of steering clear of all things strange. The only way to decide that will be to look at the range of responses that ancestral cognitive developmental structures made available and the responses that would be optimal from the perspective of overall cognitive economy. This will be a practical impossibility unless we are already in possession of rich information about how our minds are organized and how they develop. Even if adaptive thinking has some value in the generation of hypotheses to test, it will not supplant traditional work in the natural and social sciences that is essential both for filling out the background assumptions that adaptive thinking requires and for testing the hypotheses that it produces.

I am not arguing that functional approaches can have no value in illuminating mechanisms.[4] Some forms of functional reasoning, which are often mistaken for adaptive thinking, do not consist in predicting mechanical solutions from environmental problems. Instead they begin by noting that an organism does, in fact, achieve certain tasks and then proceed to posit mechanisms that might explain how this task might be achieved.[5] Obviously it is important to remember that at every stage of development some set of mechanisms must maintain an organism in a viable state (Bateson 1987). Functional analysis—the explanation of some complex capacity in terms of the contributions of functional subunits—can then help to illuminate just how survival, and the component processes of survival, are accomplished throughout the organism's life. This kind of approach is already present in the cognitive sciences where functional analysis has a long tradition. Functional analysis and adaptive thinking are not the same thing.

8.5 Case Study: Child Abuse

In Daly and Wilson's work on violence (particularly homicide) within the family, the story of how evolutionary psychology has been used to make novel, well-confirmed predictions seems strong. Daly and Wilson note that considerations of how our ancestors might be expected to have maximized their fitness could lead us to suspect that a parent that mates with an individual who already has infants will have little interest in making great investments in those infants, because they will not make any contribution to the parent's fitness. Hence, we should expect, on evolutionary grounds, considerable conflict between step-parents and their stepchildren, in comparison to conflict between parents and their biological children. Daly and Wilson were surprised to see that no research had been undertaken to test the hypothesis that abuse is higher between step-parents and stepchildren than between biological parents and their biological children. Research they conducted into abuse in Canada and the United States yielded the conclusion that: 'stepparenthood *per se* remains the single

---

[4] I am grateful to Patrick Bateson for persuading me to take into account the successes of the functional approach.
[5] This is what I have elsewhere called 'Weak Adaptive Thinking' (Lewens 2002).

most powerful risk factor for child abuse that has yet been identified' (Daly and Wilson 1988: 87–8).

So here we seem to have a straightforward case where evolutionary psychology has illuminated a previously unrecognized fact, and one that may well have important policy implications. On reflection, however, we can challenge both the importance of evolutionary thinking in uncovering the significance of step-parenthood and the policy recommendations that flow from the recognition that step-parenthood is an important risk factor for child abuse.

First, is it true that evolutionary thinking leads directly to the prediction that male parents, for example, should harm stepchildren? In an intelligent species, where the female is able to refuse to mate and also able to detect the behaviour of her mate, one might suppose that in order to maximize her own reproductive fitness she would withhold mating opportunities from any male whom she knew or suspected of abusing her own offspring. Hence, in such a species, one might expect, also on evolutionary grounds, that the maximization of female fitness would lead to a set of evolutionary problems in which males would be driven to care for stepchildren as their own. Daly and Wilson were aware at the time they formulated their predictions of a large body of work among social workers that suggested high rates of abuse among stepfamilies before they came to do their own empirical research, and one might suspect that but for this knowledge, the question of what evolutionary thinking most naturally predicts is left open.

Second, one might also ask whether evolutionary thinking is required to generate the hypothesis that abuse will be greatest in stepfamilies. Folk psychology also gives us an engine for the generation of hypotheses, and, like evolutionary thinking, some of these hypotheses will turn out to be correct. Folk psychological reasoning could certainly have generated Daly and Wilson's hypothesis that, in general, levels of antagonism will be greater between step-parents and their stepchildren than between parents and genetic children.[6] In stepfamilies one or both partners may have suffered divorce or bereavement, and increased levels of tension or conflict between step-parents and stepchildren might be anticipated. Step-parents may also experience problems fitting in to a family regime

---

[6] Kitcher (1985) does not discuss Daly and Wilson's work; however, the line of argument I pursue here owes a lot to the general form of some of Kitcher's arguments.

that has its own entrenched rules and customs, built up over a number of years, in comparison with those parents who are involved directly in the construction of such rules and customs. I do not mean to suggest that folk psychology guarantees the truth of this prediction—what folk psychology leads us to test may turn out to be false. But remember that the claim we are assessing here is that adaptive thinking gives us an especially strong procedure for the generation of hypotheses, and even evolutionary psychologists agree that such predictions on the basis of evolutionary expectations require demanding tests. Daly and Wilson themselves explain the apparently low rates of abuse between adopted children and their parents by a folk-psychological explanation of sorts. They conjecture that the screening methods employed by adoption agencies bias the sample of non-genetic parents towards a minority who are better placed than most to provide stable and loving homes for non-genetic children.

Daly and Wilson withhold judgement on the question of whether genetic relatedness directly determines likely levels of abuse or whether a parent who spends time with a non-genetic child from some time around the birth of the child enjoys the same kind of low risk of subsequent abuse as a genetic parent. Adaptive thinking seems neutral between both hypotheses, yet from the perspective of policy rather a lot turns on it. Daly and Wilson claim that we have adaptations for directing love and other forms of parental investment to genetically related children. If, in the evolutionary past, children that remained near to the parent for a sustained period from birth or soon after birth were nearly always genetically related to the parent, then there would have been no need for the parent to have developed a mechanism that determines directly whether a child is genetically related. Under such circumstances, cues for the generation of love based on length of relationship and proximity will suffice to ensure that, as a matter of fact, only genetically related children are given large parental investments. For this reason, Daly and Wilson's work is compatible with a traditional view that love for a child is built up over a sustained period and relies on a complex shared history, whether or not the child is genetically related. So the policy implications of Daly and Wilson's work are far from clear. They certainly do not show that we should discourage adoption or foster parenting; as we saw, Daly and Wilson suggest that many adoptive parents may be able to take perfectly good care of a child in spite of the lack of genetic relatedness. Their work does, however, suggest that newly formed relationships between parents and children are likely to be less stable than more long-standing ones.

What Daly and Wilson object to most strongly is a particular view of parenthood and parenting behaviour which they call 'role theory'. It is in overturning role theory that they think their work has greatest value. Here is how they describe this view:

> Parenthood is typically considered a 'role'. This metaphor [...] is positively misleading [...] in its implication of arbitrary substitutability. A role is something that any competent actor who has studied the part can step into, whereas parent–offspring bonds are individualized, and cannot be established at will. (Daly and Wilson 1988: 92–3)

Tensions surrounding step-parenting are then attributed, apparently, to the 'newness' of these roles. It is because the role of step-parent is new that it is also not fully articulated, and tension arises because of the ambiguities in the role's expectations. However, Daly and Wilson complain:

> Stepparents do not find their roles less satisfying and more conflictual than natural parents because they don't know what they are supposed to do. Their problem is that they don't want to do what they feel obliged to do, namely, make a substantial investment of 'parental' effort without receiving the usual emotional rewards. (1988: 93)

There are problems in this chain of thinking. First, if 'role theory' really is as Daly and Wilson say, then they are right to think it is silly. Parenting is a skill, and it can no more be learned automatically by someone with the right intentions than bicycling or karate. What is more, one cannot simply decide to love another person, and one certainly cannot decide to love another person in a peaceful and mutually supporting way. I very much hope that 'role theory' asserts none of these things, but if it does, then evolutionary psychology is not needed to point out its failings.

Second, we should remember that everything Daly and Wilson say about the adaptive basis for differences in abuse between step-parents and stepchildren and biological parents and biological children is compatible with the claim that love and commitment between a parent and child are best built over time, and that when time is short this love may be weaker. When we imagine a step-parent in a difficult, new relationship with a child, we can describe the situation as one where the newness of the relationship is difficult for the step-parent to fit in to and where lack of a shared past can result in difficulties between the two parties. Alternatively, we can describe this as a situation in which the parent does 'not want [...] to make a substantial investment of "parental" effort without receiving the usual emotional rewards'. The second way of phrasing the situation implies a lack of desire on the part of the step-parent to make the relationship work; however, nothing in Daly and Wilson's evolutionary hypothesis supports this implication.

Third, the plausible claim that there is an evolutionary explanation for why we have altruistic tendencies towards our children is elided with what (I assume) is a proximal explanation for why those tendencies sometimes fail. We are told: 'without recourse to the concept of evolutionary adaptation, we could not hope to understand why parental love and altruism even exist, let alone why they sometimes go wrong' (Daly and Wilson 1988: 93).

Put together in this way, the second claim is simply a non sequitur. Compare: 'Without recourse to the concept of evolutionary adaptation, we could not hope to understand why bird wings even exist, let alone why they sometimes fail.' The first part of this claim is most likely true; however, the failure of bird wings is perfectly easy to understand without the concept of adaptation.[7]

Finally, we must remember, as always, that the opposition of sociological and evolutionary psychological theories of the origins and maintenance of tension within families may be wrong-headed. Suppose, for example, that it is a deeply entrenched set of expectations and values about the 'ownership' of children, about the humiliation of cuckoldry, about jealousy, and so forth that, by permeating the developmental environments of boys and men, leads them to grow up to resent their stepchildren. Perhaps these attitudes themselves explain the transgenerational stability of increased violence towards stepchildren. In this kind of scenario, sociologists would be right to say that the causes of increased abuse in stepfamilies are primarily social, and evolutionary psychologists might also be right in saying that it expresses some adaptation, albeit one

---

[7] Some might argue that the very notion of failure or malfunction requires an appeal to the concept of adaptation. My claim here is only that we can understand how a wing has failed without the concept of adaptation—not that we can understand what it is for a wing to fail.

whose reliable reproduction across generations is mediated by learning and social structures themselves, as well as by genes.

8.6 Case Study: Rape

In their book *A Natural History of Rape*, Thornhill and Palmer (2000) explain why they think their approach will not only give us a historical explanation for the existence of rape, but will also tell us how we can best control it. Again, their reasoning is that an understanding of the evolutionary origins of rape will help us identify the underlying mechanism that produces it. Once this mechanism is characterized and understood, then we will see how best to ensure that it does not fire. We will understand what inputs lead to rape and what inputs bypass the system.

Let us note, first, that it is a mistake to think that ultimate explanations offer a deeper understanding—and hence are more valuable—than proximate explanations. Again, I do not mean to suggest that evolutionary psychologists aren't aware of this error; they are.[8] Ultimate explanations—roughly, explanations that cite the evolutionary history or current reasons for maintenance of some trait, rather than the development and mechanism underlying that trait—need not, in spite of their name, offer us a deeper understanding. Sometimes mechanisms and development are of primary interest. This is particularly the case when alteration of the trait in question is what is at stake.

There is also a danger when discussing cognitive traits to slide from ultimate to proximate explanation. This is because of the ambiguity inherent in claims about the 'reasons' for certain behaviours. Such reasons can either refer to the cognitive mechanisms that motivate behaviour or to the effects of behaviour that lead to the trait's proliferation. To take an example, it is perfectly consistent to claim that rape is an adaptation and that its presence is to be explained in terms of how it has afforded reproductive opportunities to males who otherwise would not come by them, while at the same time arguing that the motivations that lie behind rape are non-sexual and are instead associated with the desire, conscious or otherwise, to inflict violence on women. What might seem a deep conflict between the two explanations—in that the

---

[8] It is a comment made repeatedly by John Alcock in his (2001).

first claims that rape is primarily 'about' sex, while the second that rape is primarily 'about' violence—evaporates when we see that the two positions offer different kinds of explanation that do not conflict at all (Buller 1999).

Thornhill and Palmer are aware of the body of work in sociology that portrays rape as an expression of violence, rather than of sexual interest. What is more, they think this work is mistaken, even that it flies in the face of abundant evidence. I do not want to argue over who is right. Perhaps both views are right. That certainly seems a conceptual possibility if, for example, the choice whether to rape is motivated by violent urges, while the choice over whom to rape follows sexual preferences. I want to stress that even if we can substantiate the claim that rape is a male adaptation for securing reproductive opportunities for low-status males, this does not tell us anything about the truth of the sociologists' hypothesis about proximate mechanism (here I follow Sober 1993). Perhaps Thornhill and Palmer are right to claim that motivation is primarily sexual. Yet it is not adaptive thinking but the analysis of cognitive and sociological data that tells them this.

In fact, Thornhill and Palmer do not think that there is sufficient evidence to show that rape is an adaptation, rather than a by-product of other adaptive traits (2000: 84). What they mean by this is that we cannot be sure that rape is produced by a mental organ specifically shaped for the role of directing rape. Taken on its own, the hypothesis that rape is produced by mental modules that are themselves designed for other biological functions leaves open almost all hypotheses about the proximate causes of rape. Rape could be the result in part of the action of violent impulses that themselves are designed for their role in competition with other men. Alternatively, it could be produced by men's normal sexual desires interacting with additional modules that regulate the desire to harm, or some such. Rape might turn out to be of a similar status to that which the evolutionary psychologist may attribute to bestiality—not itself of any adaptive benefit, yet produced by some suite of mental traits, presumably including sexual desire, each of which is amenable to some further functional explanation. Uncovering just what the proximate mechanisms are that produce rape, and how they will interact with each other under diverse circumstances, relies primarily on detailed analysis of developmental, sociological, and other forms of data.

Even when we consider some specific version of the view that rape is an adaptation, the policy recommendations that flow are hardly revolutionary. One hypothesis already mentioned is that rape is a facultative adaptation that enables men with low status, who otherwise would have few reproductive opportunities, to get access to females (Thornhill and Palmer 2000: 67). This evolutionary hypothesis suggests at most that we can reduce the incidence of rape by enacting measures to decrease differences in status between males. This is of course compatible with the commonly held idea that the alleviation of disparities in wealth may be important in the reduction of crime. Most of Thornhill and Palmer's additional policy recommendations come from the thesis that the primary proximal motivation for rape is sexual desire, which (as we have seen) is independent of any adaptive hypothesis for the origin of rape.

It is also worth pointing out that even if there is some single mental module specialized, or simply co-opted, for rape, the further claim that regulating the inputs and outputs of this module provides the best way to control rape remains open to question. This is one of the primary motivators for the evolutionary approach: 'With respect to rape, the power of the evolutionary approach lies in its ability to identify environmental changes that may remove cues that activate the evolved mechanisms that underlie rape behavior' (Thornhill and Palmer 2000: 153).

The sexual behaviour of the Yanomanö provides a good example of the limitations in this thinking. (The example is borrowed from Buller 1999: 103–4, who turns it to a quite different purpose.) In the Yanomanö culture, sex for women is taboo during the period between the discovery of her pregnancy and the completion of weaning. Hence, some couples will kill their own children so as to recommence having sex sooner. Nothing in this story contradicts the general evolutionary framework. We can imagine that members of the Yanomanö, like us, have a species-typical module that regulates parental love and which promotes abusive behaviours when cues informing about genetic relatedness are not present, as well as a species-typical module that regulates sex drive. However, the case provides us with an example where the killing of children is brought about not by altering the inputs to the module regulating parental love, but instead by placing a cultural barrier in front of a quite different drive—the drive to have sex—so that the drive to have sex itself results in infanticide. This reminds us that the easiest ways to control undesirable behaviours—rape, murder, theft—may not be to regulate the inputs to putative adaptations specifically responsible for rape, murder, or theft, but instead to appeal to quite different drives, such as the desire for self-expression or wealth. In other words, if the Yanomanö can be induced to kill their children through a cultural taboo that makes this the only way they can have sex, we need not dismiss the thought that potential rapists can be dissuaded from rape by imposing quite standard penalties such as incarceration. If cognitive adaptations controlling sex drive can push someone to infanticide, then we have no reason to doubt that measures such as fines and imprisonment that appeal to putative cognitive adaptations controlling economic prudence, or fear of confinement, can dissuade someone from crime. Of course one may respond by claiming that these adaptations, too, should properly be included under the 'evolved mechanisms that underlie rape behavior', for all of them have effects on rape behaviour and they are, perhaps, all evolved. The problem with this response is that it removes evolutionary psychology's promise of giving a novel set of recommendations for policy, by conceding that behaviours may be influenced in complex ways by all kinds of mental modules acting in concert.

8.7 A Last Word on the Darwin Wars

The debate surrounding evolutionary psychology is not a friendly one. Participants are quick to divide into those who are for or against the programme, and the two sides subject each other to the most bitter attacks. If the arguments of this chapter are correct, then this should not be the case. This bitterness suggests strongly that participants on both sides see evolutionary psychology as having some deep political import that must be either supported or combated, or perhaps that evolutionary psychology threatens the livelihood of non-evolutionary scientists.

What we now see is that evolutionary psychology carries no serious threat so long as it is properly understood. It carries no serious threat to the livelihood and integrity of the social sciences, for even if it is true that our minds contain a set of evolved adaptations, these adaptations will nevertheless be best characterized by undertaking continued research in sociology, cognitive psychology, developmental biology, and economics in much the way we have done. The predictive power of evolutionary thinking is weak enough that good evolutionary psychology must be deferential to the empirical output of these sciences. Nor does the provision of evolutionary explanations for such traits constitute a great political threat. If one could show that rape, for example, were an adaptation, then nothing of political import would follow. In sum, the prospects for evolutionary policy are slim.

9
What Are 'Natural Inequalities'?

9.1 The Natural and the Social

Biologists, and philosophers of biology, routinely express scepticism concerning efforts to distinguish between the natural and the social.[1] Here are a few good reasons underpinning their scepticism:

1. Traits that one might think are part of 'human nature' can also be developmentally affected by socialization. Kim Sterelny (e.g. Sterelny 2003) has argued that many traits that are routinely claimed to be 'innate', including elements of human moral psychology and human folk psychology, in fact develop reliably in virtue of an information-rich environment, which has itself been built by human groups' efforts to structure and facilitate the learning opportunities of their offspring. If Sterelny is right, then such traits are prime examples of elements of human nature that have social ontogenies.

2. Even when socialization does not directly affect development, the traits an individual develops can be indirectly affected by socially controllable resources. For example, it has long been suggested that breast-feeding increases IQ (Anderson et al. 1999). One might think of breast milk as a 'natural' resource. Yet the ability of mothers to breast-feed successfully can be influenced by the social support made available to them in the early stages of motherhood.

---

[1] This chapter was first published in *Philosophical Quarterly* 60 (2010): pp. 264–85. Earlier versions were presented at the ISHPSSB conference in Vienna, at the IHPST (Paris), and at Bristol University. I am grateful to the audiences there, and also to Stephen John, Serena Olsaretti, and an anonymous referee for helpful comments on earlier drafts. Financial support came from the Isaac Newton Trust and the Leverhulme Trust. I am also grateful to the IHPST (Paris), where this work was completed.

3. The question of whether some physiological trait is a good one to have, including the apparently biological question of what its normal biological function is and whether it can perform that function successfully, is in part dictated by the social environment in which the person who has it lives. Dyslexia, for example, appears to have a neurological basis, but the inability to read and write effectively is no handicap in pre-literate societies.
4. Biological adaptations can reside outside an animal's skin (Dawkins 1982). Beaver dams are best understood as adaptations of beavers. They are at one and the same time parts of beaver societies and parts of beaver nature. Once this is acknowledged for beavers, it is hard to deny that human institutions such as schools are parts of human nature and also parts of human societies. Hence, access to schooling might be thought of both as access to a natural resource and as access to a social resource.

While scepticism of the natural/social distinction is rife within philosophy of biology, a version of the natural/social distinction plays an important role in some areas of political philosophy. In this chapter I shall largely restrict my discussion to the work of some prominent Rawlsians, and leave open the issue of whether related issues arise in the work of others (e.g. in Dworkin's treatment of 'genetic luck', 1981: 313). The background idea that makes the distinction salient is as old as Plato. Sometimes people are worse off than others through bad luck alone. When inequality is nobody's fault, it is not the concern of justice to correct it. It is tempting to rephrase this background intuition by invoking the natural/social divide: inequalities that derive from the 'natural lottery' are misfortunes; only inequalities that derive from social causes can be thought injustices. As Rawls puts it, 'The natural distribution is neither just nor unjust' (Rawls 1999: 87). Of course, the claim that natural inequality lies wholly outside the realm of justice is a particularly stark version of the conviction that misfortune and injustice should be treated differently, but several prominent theorists have argued that there is an important difference between natural and social inequalities. My question is one which is little discussed in any detail (except by Rosenberg 2000 and Lippert-Rasmussen 2004): can the distinction these theorists rely on be clearly drawn?

The structure of the chapter is as follows. I begin with a very brief description of influential work by Buchanan and co-workers, in their book *From Chance to Choice* (Buchanan et al. 2000). They endorse a distinction between the natural and the social. I then give an equally brief explanation of why one might fear that there is no such distinction to be had. Having set up the opposing views in caricatured fashion, I then look at one recent attempt on the part of a moral philosopher (Lippert-Rasmussen 2004) to draw the natural/social distinction, which uses the Analysis of Variance (ANOVA), before looking in more detail at proposals from Buchanan et al. and Nagel (1997). I argue that the feasibility of social control offers the most plausible way of making sense of the distinction. While there is a genuine distinction to be had here, it is misleadingly described. 'Natural' and 'social' are inappropriate ways of *labelling* the inequalities in question. Furthermore, there are reasons for thinking that the natural/social distinction, when drawn in this way, is of questionable relevance to theories of justice.

9.2 Natural and Social Inequalities

Is it really true that political philosophers rest weight on the distinction between natural and social inequalities? Thomas Pogge (1989) thinks that John Rawls endorses such a distinction (Rawls 1971). Pogge asks the important question:

> Faced with some given prospect of being less well educated than others, why should one care whether it is because of social factors (official prohibitions or high tuition fees) or a natural handicap (blindness, say) that is not evened out through social institutions? (Pogge 1989: 56)

He gives us an answer on Rawls's behalf:

> It could be said that Rawls's attempt to secure an acceptable share of social primary goods for every participant treats some participants, namely those who suffer special natural handicaps, unfairly. Rawls can reply [...] that it is not the role of an institutional scheme to even things out in the interest of the overall justice of the human universe (the institutional scheme included). What persons may reasonably demand of an institutional scheme is only that it should situate them fairly as participants vis-à-vis the others. Society's response to the blind objector would then be that, by hypothesis, the amount of resources devoted to his education represents a share that would be fair for a normal, sighted participant. His natural handicap, through no fault of his own, is in no way a

consequence of social institutions (or any other social factors, for that matter). Therefore the claim for additional resources he addresses to his fellow participants can appeal only to morality, not to justice. (Pogge 1989: 56–7)

Here, the natural/social distinction is vital. So long as the blind person gets a fair share of social resources, justice is done. There is no further need to compensate for a natural handicap, precisely because such a handicap, while generating inequality, is not a consequence of any social institution.

Before proceeding, I should immediately make two concessive remarks. First, much of Pogge's recent work falls closely into line with the sceptical conclusions of this chapter. As he puts it, 'Natural and social factors interpenetrate in their effects [. . .] And natural factors are by no means independent of social factors' (Pogge 2004: 155). I cite Pogge's earlier work here to show only that the natural/social distinction has had currency in prominent works of political philosophy. Second, I shall not take a stand on precisely what Rawls's own views may have been on these matters. For example, in response to critics who argued that he was wrong to focus exclusively on the distribution of social goods, Rawls says that in *A Theory of Justice* he was assuming for the sake of simplicity that all citizens had adequate levels of mental and physical health for the duration of their lifetimes (Rawls 1993: 183). It is not clear how Rawls's stance should be interpreted once this assumption is given up.

Thomas Nagel (1997) also gives salience to the distinction between natural and social inequalities. As Nagel sets up the problem, it applies to 'at least some of the inequalities that arise between men and women, or between persons differently favoured by what Rawls calls the natural lottery, or between those who are more and less energetic or enterprising—through the operation of a system of social and economic relations that is not designed to produce those results but that produces them nonetheless' (Nagel 1997: 304).

Nagel's thought is that 'since the interpersonal differences that produce the inequalities are not socially created but natural, the responsibility of society for avoiding such results is not clear' (1997: 305). As he later puts it: 'it seems to me morally intelligible to hold that because it is nature that has dealt them this blow, a social system that does not engage in significant rectification of the inequality is not guilty of injustice' (p. 315). He credits Rawls with the view that even if natural inequalities make some demand on justice, natural inequalities do not make the same demands as social inequalities. One may think this is surprising: after all, Rawls at times appears to invoke the natural lottery precisely in order to argue that it is just as unfair as the social lottery. In Nagel's words, Rawls's idea appears to be: 'What can be said of being born with a silver spoon in your mouth also goes for being born with golden genes' (1997: 309). Part of Nagel's reason for attributing to Rawls the view that social inequalities make more significant demands than natural inequalities comes from Rawls's insistence that the principle of fair equality of opportunity is lexically prior to the difference principle. Nagel asks, 'Why should the prohibition of exclusionary discrimination in employment take priority over the difference principle, which is designed to combat the unequal effects of the natural lottery?' The question is a good one, for Rawls might have taken a different turn. Those who have been blessed by the natural lottery might have been deliberately excluded from favoured jobs in an effort to equalize overall shares. The fact that Rawls prohibits this by introducing lexical priority of the fair equality of opportunity principle suggests that for him there is something worse about social discrimination against those with superior talent than there is about a failure to ensure that those with different natural talents end up equally well off.

A version of the distinction between natural and social inequalities is especially important in the discussion of genes and justice in Buchanan et al.'s *From Chance to Choice* (2000). The authors of this book are not blind to the blurring of the social and the natural. They point out both that a trait is valuable only in certain social surroundings (p. 79) and that whether some trait is under social control changes as technology changes (p. 83). They also understand the important 'norm of reaction' concept, which I make extensive use of later in this chapter. Finally, Elliott Sober's excellent appendix to the book (Sober 2000) draws attention to many of the pitfalls involved in distinguishing nature from society. Even so, their invocation of the distinction between the natural and the social remains problematic.

In chapter 3 of their book, Buchanan et al. distinguish various senses of equality of opportunity, and question how such concepts might be applied to genetic justice. One of these they name the 'level playing field' concept, borrowing the phrase from Roemer (1996):

Equal opportunity requires not only the elimination of legal and informal barriers of discrimination, but also efforts to eliminate the effects of bad luck in the social lottery on the opportunities of those with similar talents and abilities.
(Buchanan et al. 2000: 65)

They go on to distinguish two variants of this conception: the 'brute luck' view (here following the terminology of Scanlon 1989) and the 'social structural view'. The brute luck view has it that all contributions to unequal opportunity should be overcome, regardless of whether they derive from the arbitrariness of the 'natural lottery' or the 'social lottery'. The brute luck view allows that we might compensate those who have a poor genetic endowment by giving them compensatory schooling, or nutrition, or even (in the future) by giving them a different set of genes. Buchanan et al. say that this view is held by Roemer (1996), Arneson (1989), and Cohen (1993).

The social structural view, which Buchanan et al. back, differs from the brute luck view, specifically with respect to its treatment of natural inequality:

The social structural view [...] limits the domain of equal opportunity in the first instance, to social inequalities, because it is concerned only with how social structures, and more specifically unjust social institutions, influence a person's success in competing for desirable offices and positions in society. The brute luck view is much more expansive: It enlarges the domain of equal opportunity to include natural inequalities. (2000: 67–8)

One needs to be careful here in saying just how the social structural view circumscribes the equality of opportunity. The important contrast is not between inequalities owed to social and those owed to natural lotteries. Rather, the contrast is between inequalities owed to unjust institutions and those owed to misfortune.

The social structural view does not entail that *all* social inequalities should be corrected, for they may not all be unjust. On the face of things, Buchanan et al. might be taken to argue that we should correct *only* social inequalities. That is, one might take them to endorse the view that while some social inequalities may be unjust, no natural inequalities are unjust. In fact, even this goes beyond what they claim, for they subscribe to Norman Daniels's view that the importance of healthcare lies in equal opportunity and that it requires us to prioritize the ability of all to become 'normal competitors' (Daniels 1985). When our natural endowments fall below this baseline, equal opportunity requires us to be raised to it. But this does not require all natural inequalities to be equalized in the name of equality of opportunity, for some natural inequalities may exist between individuals who are already well above the relevant baseline. It does require all social inequalities that are generated by unjust institutions to be equalized. On the social structural view, our duties with respect to natural inequalities remain more limited than our duties with respect to social inequalities. Once again, the underlying idea is that all natural inequalities are the result of bad luck, but some social inequalities are the result of injustice.

9.3 A First Failed Way to Draw the Distinction

First, here is a particularly bad way of distinguishing between natural and social inequalities. One might think that it can be drawn in simple causal terms. Natural inequalities are caused by differences in natural resources, while social inequalities are caused by differences in social resources. The first problem this proposal faces is how to decide which resources are natural and which social. As I have remarked, apparently 'biological' factors which contribute causally to the development of valuable traits, such as the presence of breast-milk at birth, can also be affected by social circumstances. Moreover, apparently 'social' factors, such as the availability of clothing and shelter, might be thought of as part of our species' natural endowment.

The view that natural inequalities are caused by differences in natural resources faces a second problem, which I will focus on for most of this chapter. Suppose we can provide a good account of which causal contributors to inequality are natural and which are social. Suppose we consider genes as canonical natural resources, and suppose we can also find some canonical set of social factors that contribute to the development of valuable human traits. How are we to decide which inequalities are caused by genetic differences and which are caused by social differences, given that most valuable traits will be produced by both types of factor interacting together? It is a platitude of behavioural genetics that the causal action of a given gene is dependent on the environment in which it is found. Moreover, different genotypes can respond quite differently in different environments. To give just two examples of behavioural genetic studies of gene–environment interactions, a widely reported study by Caspi et al. (2002) examined the causal impact of abuse in childhood on later antisocial behaviour in men. They concluded that the influence of childhood abuse was partly contingent on the presence of a genetic variant that appears to reduce the activity of the enzyme monoamine oxidase A (MAOA). Childhood abuse appears particularly likely to promote aggressive behaviour in men with this variant. The MAOA genotype was not found to be associated with aggressive behaviour in men who had not been abused. A more recent paper by Caspi et al. (2007) looked at the apparent positive effects of breast-feeding on IQ. They concluded that a variant of the FADS2 gene affected an infant's ability to profit from breast-feeding: babies with the variant in question did not enjoy any appreciable IQ boost. Both studies confirm the general thought that a given trait—aggressiveness, IQ—can be affected by the social environment—childhood abuse, access to breast-milk—but that the question of what impact these social factors have is itself contingent on the presence of genetic variants.

Clearly this renders problematic any attempt to distinguish effects of natural differences from effects of social differences. To be more specific, it makes problematic the way in which political philosophers sometimes speak of 'natural endowments', using the language of 'genetic predisposition'. In spite of his scepticism about crude ways of drawing the natural/social distinction, Pogge says: 'Someone genetically predisposed toward good health, cheerfulness, intelligence, and good looks is better able to advance her conception of the good than someone who is genetically predisposed toward sickliness, melancholy, low intelligence, and ugliness' (Pogge 2004: 152). But if the causal effect of an individual's genes is contingent on the social environment, we cannot assume that the direction in which those genes 'predispose' that individual's development is wholly in the hands of nature.

In order to sharpen the conceptual issues at stake in distinguishing natural and social inequalities, we will focus on potential ways in which genetic and social environmental factors might contribute to the development of a valuable trait such as intelligence. The discussion which follows ignores various significant complicating factors, such as the difficulties of assuming that intelligence is a single general quality. I also take no stand on the issue of whether intelligence is measured

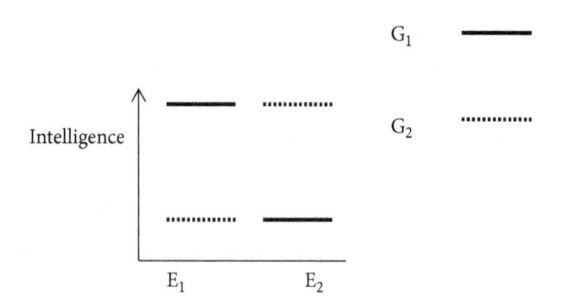

Fig. 9.1 Hypothetical norms of reaction for intelligence

by IQ. I make use of the 'norm of reaction' concept. Norms of reaction illustrate by simple graphs the responses of different genotypes in different developmental environments. Suppose, then, that genotypes G_1 and G_2 have the norms of reaction expressed in Fig. 9.1. G_1 yields very high intelligence when it develops in environment E_1, but low intelligence when it develops in environment E_2. For genotype G_2, the situation is reversed. E_2 is in fact very rare, but also easy to make widely available, were we to choose to do so. E_2 is a peculiar educational environment that suits the genes of people with G_2. For the moment, these hypothetical norms of reaction suffice to illustrate the possible difficulties we might have in demarcating natural from social inequalities. I consider in Section 9.5 the worry that they are too unrealistic to underpin any significant conclusions regarding the natural and the social. Caspi et al.'s 2007 paper covers a real case in which quantitative geneticists have argued that the positive effects on IQ of a particular environment are contingent on the presence of a genetic variant.

Suppose a society is set up so that a child, A, with G_2 grows up in E_1. The difference in E_1 between A and another child, B, with G_1 is certainly owed to genetic differences. So we might say that it is a natural inequality, because we think of natural inequalities as inequalities caused by genetic differences. But the difference is also owed to society: had A had access to a different educational regime, then he would have been just as intelligent as B. So now we might say the difference is a social inequality, too. Indeed, there is no principled answer to the question of whether such a person is a victim of a *natural lottery* because he has the wrong genotype for that environment or a victim of a *social lottery* because he is in the wrong environment for his genes. His genes simply are not a good match for his environment. Of course this does not mean that the inequality in question is also the result of an unjust social institution: perhaps there can be blameless social misfortune just as much as there can be blameless natural misfortune. But once social influence is admitted in the context of some inequality, it is hard to shake off questions of justice. If monolithic education systems that ignore the different needs of individuals with different genetic endowments are unjust social institutions, then the fact that this difference in intelligence looks like a prime candidate for being a natural inequality does not by itself rule out the possibility that it is also an inequality that is the result of an unjust social institution.

This line of thought threatens to undermine the distinction on which Buchanan et al.'s social structural view relies. Suppose, in a society in which all individuals develop in a single educational environment (E_1), a small handful of individuals (G_1) are super-geniuses, and the rest (G_2) merely of moderate intelligence. Everyone has what it takes to reach the baseline of being a 'normal competitor'. But if we were to introduce a diverse educational system, in which G_1 people continue to develop in E_1, but G_2 individuals now grow up in a different environment E_2, in which they do much better, then all could be super-geniuses. An illustrative set of norms of reaction is depicted in Fig. 9.2. In the original situation, in which all are in E_1, if we class as natural the inequality then present, justice demands we do nothing about it, at least not in the name of equal opportunity. If we class the inequality as the result of an unjust social institution, justice requires us to institute a variegated educational regime in the name of equality of opportunity. It does not seem that we can use a simple causal condition to demarcate natural and

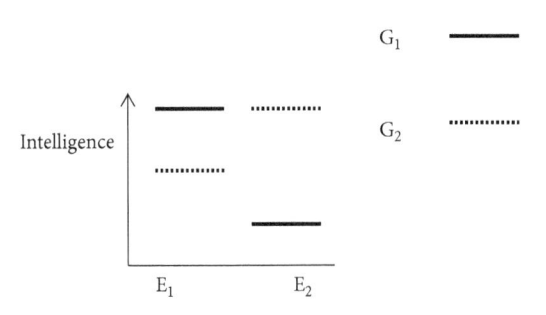

Fig. 9.2 Hypothetical reaction norms where all are 'normal competitors' in E_1

social inequalities. Buchanan et al.'s social structural view consequently requires some other way of drawing that distinction.

9.4 ANOVA Effort to Distinguish Nature from Society

Next, I shall consider a straightforward effort to demarcate natural and social inequalities, which can be found in a paper by Kasper Lippert-Rasmussen (2004). In many respects, his goals are similar to my own. He is sceptical about the distinction between natural and social inequalities. But that is because he thinks it is morally irrelevant, not because he thinks it cannot be drawn. He suggests we can make sense of the distinction by appealing to the analysis of variance (ANOVA). Only a very basic and simplified presentation of ANOVA, which I have adapted from Sober (2001), is needed to assess Lippert-Rasmussen's proposal. We take a field and divide it into plots. In some plots we put corn plants with identical genotypes and subject them to different environmental treatments. In other plots we keep the environmental treatments the same and vary the genotypes of the corn plants. We then record the average heights of the corn plants in each of the plots. Suppose the field has identical genotypes planted in rows across its width and identical environmental treatments running down its length, with average heights recorded for the plants in each plot, as illustrated in Table 9.1.

We can then calculate further averages, such as the average height of all the G_1 plants regardless of environment, the average height of all the plants in E_1 regardless of genotype, and so forth. Maybe we end up recording numbers like those illustrated in Table 9.2.

Or we might end up with results like those illustrated in Table 9.3.

Table 9.1. A cornfield, with identical genotypes planted in rows across its width and identical environmental treatments running down its length

	E_1	E_2	E_3
G_1	$H_{(G_1, E_1)}$	$H_{(G_1, E_2)}$	$H_{(G_1, E_3)}$
G_2	$H_{(G_2, E_1)}$	$H_{(G_2, E_2)}$	$H_{(G_2, E_3)}$
G_3	$H_{(G_3, E_1)}$	$H_{(G_3, E_2)}$	$H_{(G_3, E_3)}$

Table 9.2. An imaginary data set, in which 100 per cent of variance is genetic

	E_1	E_2	E_3	$H_{(AV)}$
G_1	10	10	10	10
G_2	5	5	5	5
G_3	3	3	3	3
$H_{(AV)}$	6	6	6	

Table 9.3. An imaginary data set, in which 100 per cent of variance is environmental

	E_1	E_2	E_3	$H_{(AV)}$
G_1	10	5	3	6
G_2	10	5	3	6
G_3	10	5	3	6
$H_{(AV)}$	10	5	3	

ANOVA apparently promises to quantify the respective causal contributions of genes and environment, or nature and nurture, to phenotypic traits. The details of exactly how ANOVA calculates these numbers are not important here. Very roughly speaking, the idea is to quantify the amount of 'spread', or variance, that one finds in the different average height values recorded across all the plots, and then to quantify the proportion of this variance that is attributable to genetic and environmental factors. This is done by quantifying the amount of spread in the average values for each genotype (i.e. the spread between the average figures in the rightmost columns of Tables 9.2 and 9.3) and the amount of spread in the average values for each environment (i.e. the spread between the numbers on the bottom rows of Tables 9.2 and 9.3). Table 9.2 depicts a situation in which 100 per cent of variance is genetic. Table 9.3 depicts a situation in which 100 per cent of variance is environmental.

There is a well-established tradition of scepticism about the ability of ANOVA to reveal causal influence, which originated in an influential paper by Lewontin (1974). In response to Lewontin, Sober agrees that it does not make sense to ask, of an individual organism, what proportion of some trait (the organism's height, say) is produced by its environment and what proportion is produced by its genes. But Sober points out that one can ask a different question, which refers to a population rather than an individual: how much of the variance in that population is explained by genetic variation, and how much is explained by environmental variation? Sober thinks that ANOVA does tell us the answer to this question. In this sense, he claims that ANOVA is a good way to quantify the contributions to some trait made by nature and nurture (see Sober 1988; 2001). This prima facie gives the political philosopher some hope for using ANOVA as a way of distinguishing natural and social inequalities. Inequality is itself a feature of a population, rather than an individual human. So, it seems, one can ask whether some inequality is explained primarily by natural factors or primarily by social factors. Lippert-Rasmussen takes up this proposal:

> An inequality between two individuals is primarily due to natural differences if and only if the total variation in the inequality between the two is primarily explained by variation in their natural differences. (2004: 198)

One potential limitation of this suggestion is that ANOVA will not make the natural/social distinction a matter of all or nothing. At the extremes there will be purely natural and purely social inequalities. Most inequalities will be partial mixes. For any view within political philosophy that requires natural inequalities to be treated differently from social inequalities, as Buchanan et al.'s social structural view appears to, it is not clear that a distinction that comes by degree in this way will be suitable.

This need not be a problem with Lippert-Rasmussen's proposal, especially not for those political theories which prioritize the remedy of social inequalities and for which a distinction that comes in degrees could be perfectly suitable. Even so, as ANOVA is usually conceived, it is a particularly bad way of drawing the distinction between natural and social inequalities. One concern is that there can be cases in which phenotypic variation is highly correlated with genetic variation, in virtue of the fact that genetic differences produce physiological differences, which then interact with a discriminatory environment. For example, different genes may cause different skin colour, and schoolteachers may then systematically ignore the needs of children with darker skin. If this happens, we shall find a very strong correlation between genetic factors and educational outcomes. We can also add to this scenario that all children are in the same sorts of school, regardless of their skin colour.

We might then conclude that there is no significant environmental variation. But the result encouraged by ANOVA—that this is a perfect natural inequality—seems to be the wrong one. At the very least, it is extremely misleading from an ethical perspective.

There are other problems with using ANOVA to ground the natural/social distinction. To return to our earlier example of a society with a monolithic education system, since there is no variation in social resources—all individuals are in the same educational regime—the ANOVA method will describe what variation there is as wholly explained by genetic variation, and so the inequality will be regarded as wholly natural. This is in spite of the fact that it seems plausible to attribute the inequality in question to a decision to institute a monolithic educational system, and as a result of that it seems perfectly sensible to say that the inequality in question has social causes, which are appropriate targets for scrutiny with respect to their justice. The problem here results from the fact that ANOVA, at least as it is usually conceived, overlooks social causation resulting from a failure to institute alternative social regimes.

Perhaps one of the reasons that Rasmussen thinks that ANOVA does offer a reasonable promise of distinguishing natural from social inequalities is that his understanding of ANOVA is quite unusual. As we have seen, ANOVA is usually understood to provide a population-specific decomposition of causes. Consider again some plots in a cornfield, as illustrated in Table 9.4.

ANOVA tells us that 100 per cent of variation here is explained by genetic differences. Now suppose the plants in the top left corner of the field (the ones with G_1 and E_1) stay where they are, with just the same treatment, but they are given some new neighbours (G_3) and some new environmental treatments (E_3), as illustrated in Table 9.5.

Nothing has happened to the plants in the top left corner, so of course their height is unaltered. But a new analysis of variance reveals that variation is no longer wholly explained by genes: it is wholly explained

Table 9.4. Plots in a cornfield, with hypothetical values for height

	E_1	E_2
G_1	10	10
G_2	5	5

Table 9.5. New neighbours are introduced to our cornfield

	E_1	E_3
G_1	10	5
G_3	10	5

Table 9.6. One of Lippert-Rasmussen's ANOVA matrices

	G_1	G_2
S_1	4	4
S_2	0	0

by environment. There is no surprise in this result: ANOVA is always tied to the explanation of variance in a particular population.

For Lippert-Rasmussen, things work differently. He explains his way of applying ANOVA by using a series of 2×2 matrices, such as the one laid out in Table 9.6.

One might think that the values represent recorded phenotypic means for G_1 and G_2 individuals in the two social environments S_1 and S_2 present in some society. In other words, one might assume that the whole matrix depicts the actual state in some society. But this is not how Lippert-Rasmussen's matrices work. He asks us to imagine that '(S_1, G_1) represents the actual state' and that 'the numbers in the matrixes represent inequalities between two groups of people'. G_1 is not a genotype; rather, it is a 'distribution of genes among members of this society'. S_1 is a distribution of social norms and expectations. Moreover, since G_1 and S_1 represent the actual state, G_2 and S_2 represent hypothetical alternatives. If changing the social environment to some non-actual S_2 would alleviate inequality, then Lippert-Rasmussen's ANOVA counts the inequality in question as social. Standard ANOVA, on the other hand, ranges over only actual variation in social environments.

This is a substantial departure from the normal ANOVA method, if only because we cannot conduct a Lippert-Rasmussen-type analysis without using hypothetical values for what the level of inequality would be like, given various non-actual distributions of genes and social norms. The question of whether some inequality is primarily social or primarily natural then turns on the question of how much proportional impact possible alterations to these social and genetic distributions would have.

Intuitively speaking, we might think of a social inequality as one that alterations to social circumstances could make better. I have suggested that the standard account of ANOVA is a bad way of capturing this intuitive notion, partly because even when natural variation is highly correlated with inequality, it might be the case that an unactualized social environment would considerably alleviate inequality. One might think that by allowing ANOVA to range over unactualized possibilities, Lippert-Rasmussen's account eludes this difficulty. Lippert-Rasmussen himself limits these unactualized possibilities to 'feasible' ones, and opts not to engage with the tricky question of how we are to determine which logical possibilities are to count as 'feasible' (2004: 200). This means that it is unclear as to whether his analysis really gets around the problem. But in any case, as Lippert-Rasmussen himself points out, on his analysis many inequalities will be classified as social even when they cannot be improved by altering social circumstances. This will be the case when all alternative admissible social circumstances would make some inequality much worse, while all alternative admissible natural circumstances would only make it a little better. There might be two situations of identical degrees of actual inequality; in one, changing the social environment would make the inequality much worse. In the other, changing the social environment would make it a little worse, while changing the genetic variation would make it much better. The first is then a largely social inequality, the latter a largely natural inequality; but it seems that there is no good reason to think the first situation is worse than the second in virtue of this. This is the source of Rasmussen's scepticism about the moral importance of the natural/social distinction, but one could just as easily take it as evidence that this version of ANOVA does not succeed in giving an account of what the natural/social distinction is.

9.5 Control: A Misleading Way to Draw the Distinction

The distinction between natural and social inequalities is widely held to be an important one for political philosophy; but I still have not identified a good way of saying what it takes for some inequality to be natural, rather than social. One rather obvious way to do this takes a lead from the consequences the distinction is said to have. If natural inequalities are those that do not fall within the domain of justice, then perhaps we should say that natural inequalities are simply those over which we have no control. This is quite explicitly put forward at one point by Buchanan et al., who suggest that maybe the right way to view the social/natural distinction is 'between what is subject to human control and what is not' (2000: 83). This manner of drawing the distinction has the attractive consequence that inequalities classed as 'natural' will shrink as our knowledge and technical abilities increase.

If the distinction is drawn in this way, one can then begin to raise the question whether since some inequalities are not subject to human control, they are indeed the result of bad luck rather than injustice. One might jump off from this observation to complain that in practice there is nothing wrong with the use made by political philosophers of the natural/social distinction. True enough, it is likely that genes give rise to intelligence via complex norms of reaction. But one might complain that in most cases these reaction norms are far too complex for us to unpick them and to subject intelligence, or athletic ability, to social control. One might go further and claim that the development of cognitive and physiological talents intuitively classed as 'natural' at present eludes our attempts to control them and will most likely continue to do so in the foreseeable future. Hence, it is legitimate to refer to inequalities in such talents as 'natural inequalities'.

The first thing to note is that there is evidence that certain social interventions really can influence intelligence. Here we enter murky territory, but if it is true that breast-feeding boosts IQ, then efforts to provide new mothers with advice and support to help them breast-feed could be effective in altering IQ levels. Some will construe this as a potential case whereby social intervention can make a difference to intelligence.

I move on to examine the proposal that controllability might be a good way to distinguish natural and social inequalities. One of the primary reasons for an inability to control the existence of an inequality in talents may be ignorance regarding how those talents develop. But the developmental resources that contribute to an inequality in, say, intelligence might include social arrangements as well as (for example) the distribution of genetic or other biological resources. Ignorance of the way in which social circumstances contribute to developmental outcomes can give rise to inequalities we cannot control. It is therefore misleading to describe the inequalities in question as 'natural' rather than 'social'. The choice of label is not a trivial matter: labelling inequalities in this way might give the false impression that there is no point in researching potential social interventions if we wish to equalize 'natural' inequalities. Of course, these observations do not entail that political philosophers are attempting to reach for a distinction where none is to be had; rather, the key point here is that the distinction that appears to underlie debates in political philosophy regarding natural and social inequality is really a debate about controllable and uncontrollable inequality, and that it would be better for the inequalities in question to be labelled as such.

The risks of misusing the natural/social label are borne out by the fact that Buchanan et al. say in passing that the disease phenylketonuria (PKU) is a 'natural deficit' (2000: 70), in spite of their belief that the mental retardation associated with it is under social control via the provision of a special diet (p. 72). Given that they regard natural inequalities as uncontrollable inequalities, shouldn't they refer to disadvantage conferred by PKU as a social inequality? Buchanan et al. are likely to respond by saying that they are not confused about the PKU case at all. They believe that possessing PKU genes—something that is not under our control at present but may become so—is a natural deficit, while any disadvantaging phenotypic trait the genes cause may or may not be a natural deficit, depending on whether it is under social control. Even this is odd, though, for it is unclear how to make sense of the suggestion that some set of genes is advantageous or disadvantageous independently of the developmental context that gives it positive or negative effects.

9.6 Control: A Grain-of-Analysis Problem

A second, more substantive concern is raised by the distinction between controllable and uncontrollable inequalities. The question is whether a distinction drawn in this way can plausibly be thought to have important consequences for a theory of justice. One might think that by understanding natural inequalities as those we have no control over, it becomes too obvious that they fall outside the scope of justice. After all, if we cannot control them, we can do nothing about them. And if there is nothing we *can* do about them, there is nothing we *should* do about them. Natural inequalities fall outside the domain of justice for the simple reason that there is no possibility that justice can equalize the differences in question.

This reaction overlooks a key distinction assumed in discussions of nature and justice, between *control* over some inequality in talent and *compensation* for some inequality in talent. We may, for example, acknowledge that we cannot intervene to rectify some difference in intelligence between two individuals. Even so, we might decide to compensate the less intelligent individual, by giving him a greater share of other resources over which we do have control. Most obviously, we might decide to augment the income of this individual, so that it approaches that of the more intelligent person.

There is a worry in connection with this distinction between our ability to rectify inequality and our ability to compensate for inequality without rectifying it. The question of whether some inequality can be rectified or merely compensated for is dependent on whether we describe inequality in fine-grained or coarse-grained terms. One might say that people born without the use of their legs suffer from inequality with respect to their ability to walk. There may be nothing we can do to help them to walk. Since this is an uncontrollable fact about them, they are victims of a natural inequality, and we can then discuss whether justice requires them to be offered compensation to make up for this natural shortcoming. On the other hand, one might say that they suffer from inequality with respect to their ability to get about. This inequality is under social control; we can offset it by installing wheelchair ramps, lifts, and so forth. In short, the question of whether installing ramps is viewed as compensation for a natural inequality or rectification of a social inequality depends on whether we choose to describe the inequality in question in a general or more specific way. But surely it is implausible to think that the demands that some state of affairs makes on justice depend on features of how we choose to describe that state. Suppose, then, that the notion of controllability is one that we can make reasonably precise. We may also agree that it is the best way of making sense of the distinction for which political philosophers are reaching when they distinguish natural and social inequalities. This grain-of-analysis problem undermines the claim that the natural/social inequality distinction has importance for theories of justice.

On a more speculative note, it seems that these reflections may necessitate a revision of what we mean by equality of opportunity. Buchanan et al. (2000: 129) mention 'the core idea that has long standing in our democratic culture—namely, that a competition for jobs or offices is fair if it tests people for their possession of the relevant capabilities, provided society has not unfairly distorted them'. To identify distortions of talents, it seems we must know what an individual's true talents are (see also Chapter 2, 'Enhancement and Human Nature'). But this is hard to do, given that a starting position at conception, or at birth, is likely to give rise to quite different talents in the mature phenotype depending on the environmental circumstances that are in play. Returning to the scenario in Fig. 9.1, there is no good reason to say that a G_2 individual who would be a genius in E_2 is nonetheless truly an imbecile, on the grounds that this is the level of performance he attains in the statistically typical environment, E_1. One could just as well say that a failure to provide that individual with access to environment E_2 itself imposes an unfair distortion of his abilities. The proponent of equality of opportunity needs to give an account of what it takes for opportunities to develop talents to be fair, and it is hard to see how one could justify an answer in terms of exposure to the same developmental environment. Fair equality of opportunity needs to be supplemented by an account of fair allocation of the developmental resources that enable babies to acquire talents.

9.7 Innateness

My discussion of the proposal that the natural/social inequality distinction should be understood in terms of the uncontrollable/controllable distinction highlighted the concern that in labelling inequalities as 'natural' we might falsely imply that it is not worth investigating social interventions that might alleviate them. A related worry arises in the context of a rather different proposal, namely, that natural inequalities concern differences in individuals' 'innate' talents, while social inequalities concern differences in socially acquired traits.

The innate/acquired distinction is at least as murky as the natural/social inequality distinction it is meant to analyse. Indeed, several philosophers and biologists have argued that innateness is a fundamentally confused notion, which we would be better off dropping altogether (e.g. Griffiths 2002; Mameli and Bateson 2006). One of the most plausible accounts of innateness proposes that an innate trait is one whose development is particularly robust: the same trait is bound to appear, regardless of precisely how the organism's developmental environment is configured. Innateness, on this view, is to be equated with 'canalization' (Ariew 1999).

The innate/acquired distinction, when spelled out in this way, is not the same as the uncontrollable/controllable distinction. An organism's height might depend in various highly non-linear ways on the amount of some nutrient it receives while a juvenile. We may be unable to control the organism's height if we are unable to regulate in a precise enough way how much of the nutrient the organism receives. But the organism's height is not innate.

To say that an innate trait is one whose development does not depend on the nature of the environment is vague. To take an extreme example, traits that develop in the same way, regardless of which *terrestrial* environment they are in, may not develop in the same way if the gravitational field in which they develop is varied by taking them into space. Innateness needs to be defined against some background specification of admissible alternative environments. But this raises the possibility that the development of an apparently 'innate' trait could be derailed if some very unusual social environment, not included in the specification of admissible alternatives, were put in place. This means that characterizing a trait as 'innate' does not mean that it is worthless to examine how alterations to the social environment might change how it develops. It means only that the trait's development is invariant across some restricted range of environments. Once again, the distinction between natural and social inequalities, if spelled out in terms of innateness, threatens to have misleading connotations.

9.8 Nagel on Nature

I shall finish with a discussion of Thomas Nagel's views on these matters. He too thinks that the claims on justice are not clear in the context of the victims of natural inequality: 'it seems to me morally intelligible to hold that because it is nature that has dealt them this blow, a social system that does not engage in significant rectification of the inequality is not guilty of injustice' (Nagel 1997: 315).

Having noted how Rawls assumes the apparently equal moral arbitrariness of both the natural and social lotteries, Nagel notes that inequality can have quite different types of cause and that one may feel that the differences between these types of cause are morally relevant:

> Discriminatory exclusion is practised intentionally by individuals and firms; class is a predictable effect of the operation of the social system; natural talent is biological. (1997: 309)

Even if we agree with Nagel that the aetiology of inequality is morally relevant to the demands it makes on justice, the specific taxonomy of inequalities suggested by this remark is cross-cutting. Many 'natural talents'—if by this we mean valuable talents with a physiological basis of some kind, such as intelligence and athletic ability—also rely on the social environment, and hence on the 'operation of the social system', for their development. If Nagel is attempting to implicitly stipulate that by 'natural talents' he means those talents over which the social system has no influence, then we can accuse him at best of a confusing way of writing. But if he means to make a substantive assertion about the independence of the biological and the social, then he is mistaken, and mistaken in a way that is likely to obscure important concerns regarding the just provision of developmental opportunities.

Nagel notes that the crucial question for anyone who wants to defend the claim that natural inequality falls outside the scope of justice is how we are to distinguish natural from social causation. But his answer presupposes a distinction between natural and social inequality:

> we have to be able to give sense to the idea that a difference in social outcomes is due primarily to a natural difference. Let us suppose that the outcome is not identical with the natural difference, and furthermore, that it could not even have appeared without the social institutions that create the dimensions of variation in which it arises. (Nagel 1997: 314)

In a sense, Nagel is interested in a different set of problems from those confronted in this chapter. He assumes that there is a recognizable class of natural differences—differences in intelligence, say—and that these interact with social institutions to produce significant inequalities, such as inequalities of wealth. He is after a formula that decides when it is appropriate to lay the responsibility for the resulting inequalities on natural differences rather than social institutions, given that both are always involved causally. He then proposes:

one possible account of what needs to be the case for the claim to be nevertheless plausible that the outcome is due primarily to a natural difference. It requires

1. that there is a variable natural property of individuals that plays a significant causal role in the generation through social institutions of outcomes which differ substantially in value for those individuals,
2. that the institutions not aim to produce the differential results but have an independent and legitimate purpose,
3. that to achieve that purpose without generating such differences would be significantly more difficult or costly. (Nagel 1997: 314)

Nagel is not attempting to confront the problems posed by the fact that social environments are involved in the development of many of the important differences between people that we might be inclined to class as 'natural'. Indeed, his proposal works best in the context of a simplifying assumption that there is no biological development and that individuals are born with some set of talents and abilities fully formed, which then interact with a more or less just social environment. In line with this, there are moments when Nagel relies on a form of stipulation to isolate his position from the worries posed by social influences on biological development. So, for example, he asks us to consider, as an example of a 'purely natural inequality', a lethal genetic condition that is incurable. We are told that the development of the disease is inevitable: it will appear no matter what efforts are made through medical science to change its developmental environment. In these circumstances there are no socially controllable resources that can affect the disease. We might then concur with Nagel: 'it seems to me morally intelligible to hold that because it is nature that has dealt them this blow, a social system that does not engage in significant rectification of the inequality is not guilty of injustice'. Pogge suggests a similar simplification is at work in Rawls's writings: 'citizens creating and upholding a social order bear responsibility for any very low social positions it may produce, but not for any very poor natural endowments, which can be viewed as so-to-speak preexisting' (Pogge 2004: 153).

As a matter of fact, natural endowments are not pre-existing, and it is essential to bear this in mind when we try to move beyond the simplifying assumption that valuable traits do not develop over time through interactions with the social environment. In the context of education policy, Nagel suggests:

WHAT ARE 'NATURAL INEQUALITIES'? 167

once society provides fair equality of opportunity, it is nature, not society, that is responsible for the unequal capacity of individuals to benefit from it. Educating individuals to the limit of their capacity is a legitimate aim, and social inequality generated in the pursuit of a legitimate aim is not unjust if natural differences among the persons involved are its primary cause. (2007: 316)

In the first sentence of this assertion, one might take Nagel to be claiming that so long as everyone has roughly the same access to educational facilities, we can blame nature for unequal capacities to benefit from it. This is a view we should certainly reject. It fails to account for the possibility that a monolithic education system, to which all have access, might strongly favour some learning styles over others. But in the second sentence, Nagel also suggests that there is nothing wrong with inequalities that arise as a result of educating individuals to the limit of their capacity. Aiming for this might demand that we institute a highly variegated system of educational facilities, which allows those with different learning styles all to attain their maximum levels of achievement.

Nagel's point is perhaps that even when all attain their maximum levels of achievement, there may still be inequalities between these levels and that such inequalities are not unjust. Perhaps not, but arguing the case would demand a substantial discussion of the important issues of 'the morality of inclusion', which Buchanan et al. (2000) should be credited for drawing to our attention. Fig. 9.3, for example, illustrates three alternative environments for two genotypes. While G_1 is common, G_2 is very rare. E_1 is cheap and it allows G_1 people to attain a very high level of intelligence, while G_2 people are dunces. In E_2, which is very

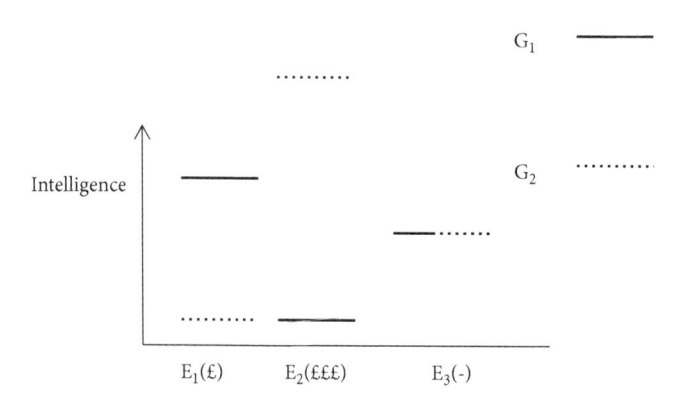

Fig. 9.3 Alternative environments, with alternative costs

expensive indeed, the situation is reversed, except that while the G_1 people are dunces, the G_2 people are super-geniuses. E_3 is even cheaper than E_1, and it allows G_2 people and G_1 people to attain a moderate level of intelligence. Any combination of these environments can be instituted simultaneously, but combinations are disproportionately expensive, because we lose economies of scale.

We would *maximize* intelligence by instituting a combination of E_1 and E_2: here the G_2 folk are super-geniuses; the G_1 folk are very intelligent; vast amounts of public money are spent on this luxurious education system, leaving little over for other priorities; and the tiny minority of G_2 people end up considerably better off than anyone else—they dominate elite institutions and are the wealthiest members of society by far. It appears that on Nagel's criteria, the resulting inequality is not unjust, on the grounds that 'social inequality generated in the pursuit of a legitimate aim is not unjust if natural differences among the persons involved are its primary cause'. His criteria for natural causation seem to imply that we should indeed attribute the resulting inequality between G_1 and G_2 people to nature, and hence place it outside the scope of justice. Against this, one can surely imagine mounting an argument in favour of the injustice of this system, compared with a uniform environment of E_3, in which all do reasonably well for very little money. The system that involves instituting E_1 and E_2 together, one might say, involves channelling an indefensibly large amount of public money towards benefiting a small minority. The point here is not to argue that this is the right way to think about justice. The point is rather that these issues seem to be highly complex, and our understanding of them is not appreciably enhanced by using a distinction between natural and social inequality.

10
Foot Note

10.1 Foot on Natural Goodness

Philippa Foot has long held that ethical judgements can be understood as factual judgements about functional performance (Foot 1961).[1] She has clarified and expanded on these views considerably in her book *Natural Goodness*. She describes her project thus: 'I want to show moral evil as "a kind of natural defect"' (2001: 5).

For Foot, moral evaluation of human action is the same in kind as the evaluation of animal and plant physiology. When we judge that some structure or process is defective, we judge it to be so with respect to functional norms for the species in question. It does not matter if we are evaluating the practical deliberations of a human or the wings of a bird: in both cases we make straightforward factual claims.

Some quick words of clarification are in order. Foot does not say that functional defects in animal or plant structures are *moral* failings. A tree's roots cannot be evil. On Foot's view, ethical evaluation applies specifically to defective practical reason, not to other defective organs or processes. But an evil person is one whose practical reason is defective, and this faculty is defective in just the same sense that a wobbly oak tree's roots are defective.

Foot's aim, then, is to provide a naturalized account of ethics by means of a naturalized theory of functions. Foot describes her account of functions as Aristotelian. On Foot's view, functional specifications consist in a list of 'Aristotelian necessities': those behaviours or traits that are

---

[1] This chapter was first published in *Analysis* 70 (2010): pp. 468–73. An early version was presented at a meeting on 'La Notion de Fonction: Des Sciences de la Vie à la Technologie' at the Collège de France, Paris, in May 2008. I am grateful to the audience there for comments and to Hallvard Lillehammer for discussion. I am also grateful to the Leverhulme Trust and the Isaac Newton Trust for funding.

necessary for the good of each 'species' or 'life-form'. Any token organism can then be judged as defective if it fails to match the functional specification for its species.

10.2 A Biological Objection to Foot

Those working in philosophy of biology may have worries about Foot's account of species and also about her account of biological functions. Here I want to focus on just one objection, which I choose because it has general force against any attempt to link moral good or evil to a naturalized account of functional performance.

For Foot, once again, function claims concern what she calls 'Aristotelian necessity [. . .] that which is necessary because and in so far as good hangs on it' (2001: 15). She gives some examples of Aristotelian necessities:

We invoke [this] idea when we say that it is necessary for plants to have water, and for lionesses to teach their cubs to kill. (p. 15)

In the case of plants and non-human animals, the species' good is fairly tightly circumscribed:

The way an individual *should be* is determined by what is needed for development, self-maintenance, and reproduction: in most species involving defence, and in some the rearing of young. (p. 33)

Foot's general case draws on an apparent affinity between normative notions of an organism's 'flourishing' and biological notions of good functioning. This link, although seductive, is misleading. The alleged link is foreshadowed in Foot's much earlier work. Writing of an oak tree's roots, she says that:

Because the root plays a part in the life of the organism we can say it 'has a function', relating what it does to the welfare of the plant. (Foot 1961: 59)

This already introduces a conception of biological function as that which contributes to the *welfare*—and not the *fitness*, note—of an organism.

We can illustrate some of the problems that face Foot's view by considering the fact that for individual organisms, there can be trade-offs between self-maintenance and reproduction. Suppose we ask how a given organism should be. Should it live for a long time, thereby using resources that would have enabled it to have more offspring? Or should it have a small number of healthy offspring and then die shortly afterwards, leaving resources to future generations? What determines whether reproduction or self-maintenance is the more important element of flourishing? Foot seems committed to the claim that natural facts alone can decide the issue.

The evolutionary biologist might say that the optimal phenotype is that which most effectively promotes long-run fitness. But it is implausible to think that what is optimal in the evolutionary biologist's sense corresponds with that which promotes the flourishing of individuals. Consider a species in which the members with the greatest reproductive output die earlier than they need to, but have more offspring in virtue of this. Other members of the species live considerably longer, but they have somewhat fewer offspring. It is implausible to think that these latter members fail to flourish simply because their reproductive output is lower than that of fitter conspecifics. True enough, some evolutionary biologists might say that they 'malfunction' in virtue of this fitness gap. Here one effectively stipulates that good functioning is to be equated with some notion of biological adaptation. But from the perspective of individual health and flourishing, these longer-lived individuals might be judged better off than those that die early. It is not clear that any naturalized theory of function can tell us, in the face of trade-offs such as these, where flourishing lies. There is a gap between the sort of normative notion of flourishing that Foot seeks to capture and theories of good biological functioning based on reproductive output.

10.3 Potential Responses

In this penultimate section I briefly consider two responses Foot might make.

10.3.1 The human good is more complex than the good of plant or animal species

Foot might remind us of some of the complexities of her view. For plants and many non-human animals, the species' good is determined by self-maintenance and reproductive output. For humans, things are more complicated. Foot says, for example: 'The goods that hang on human cooperation [. . .] are much more diverse, and much harder to delineate than are animal goods' (2001: 16). She might then try to argue that once these more complex elements of human good are factored in, the sorts of trade-offs I have laid out above can be resolved.

The primary problem with this response is that Foot's account is meant to establish that moral evaluation in the human case consists in a complexification of a fundamentally similar form of evaluation that we can bring to plants and animals. Even if the case for humans has many more dimensions than plant cases, it is important for Foot that her account should begin by making sense of the relatively simple evaluations we use when assessing parts of plants. But the problem raised by the case of trade-offs between self-maintenance and reproduction is that even in the simple cases of plants and animals, it seems that evaluations based on theories of biological functioning do not deliver plausible verdicts with respect to normative notions of the flourishing of a plant or animal.

There is a secondary problem. Foot tells us that there are more elements to the human good than reproduction and self-maintenance. She says, for example, that friendship is an element of the human good. But it is unclear what allows her to generate these aspects of the good. Aristotelian necessities, remember, concern 'that which is necessary because and in so far as good hangs on it'. A tree whose roots do not allow it to stand upright is faulty, in virtue of the fact that strong roots are necessary for the tree's good, which is constituted by its self-maintenance and reproduction. Even if one grants that we can make sense of characteristics that are necessary for the species good once that good is fixed, Foot says rather little about what makes it the case that something is an element or aspect of the good itself. This becomes an especially pressing issue once one wishes to claim that the good for humans is not exhausted by apparently 'biological' qualities such as reproduction and self-maintenance. Foot's general project here is to show that 'the norms we have been talking about so far have been explained in terms of facts about things belonging to the natural world' (2001: 36-7). But a residual worry remains regarding the factual status of aspects of the good.

10.3.2 Foot's account of function should not be confused with the technical biological notion of function

In an important note, Foot asserts:

> It is imperative that the word 'function' as used here is not confused with its use in evolutionary biology [...] It is easy to confuse these technical uses of words such as 'function' and 'good' with their everyday uses, but the meanings are distinct. (2001: 32)

Perhaps the biological example of trade-off that I discussed above only raises problems in the context of a technical definition of biological function, which equates function with contribution to fitness or some such. More specifically, Foot might remind us that her Aristotelian theory says that what underpins how species members ought to be is not that which *contributes*, or has contributed, to development, self-maintenance, and reproduction, but rather that which is *necessary* for these things. Presumably there are some activities that contribute to these ends, but members of the species would get by perfectly well without the activities in question. In the context of the example we have been discussing, Foot might observe that even if dying early contributes positively to reproductive output, it is presumably not necessary for reproduction. Hence, we need not say that a longer-lived species member is defective, for that species member might not fail in any activity that is necessary for the species' good.

This response raises far more problems than it solves. Given that evolutionary considerations suggest that adaptation occurs when one trait offers a fitness advantage over another, there will typically be times when the favoured trait is not necessary but is nonetheless beneficial. So if we really insist on the strong claim that functions are only given to traits that are necessary, our account will be extensionally inadequate. Foot's account suggests that something as strong as this is what she has in mind:

We are, let us suppose, evaluating the roots of a particular oak tree, saying perhaps that it has good roots because they are as sturdy and deep as an oak's roots should be. Had its roots been spindly and all near the surface they would have been bad roots; but as it is they are good. Oak trees need to stay upright because, unlike creeping plants, they have no possibility of life on the ground, and they are tall heavy trees. Therefore oaks need to have deep sturdy roots: there is something wrong with them if they do not, and this is how the normative proposition can be derived. The good of the oak is its individual and reproductive life cycle, and what is necessary for this is an Aristotelian necessity in its case. Since it cannot bend like a reed in the wind, an oak that is as an oak should be is one that has deep and sturdy roots. (2001: 46)

Oaks need strong roots, because strong roots are the only possible way for them to stay upright. One might quibble with Foot's notion of modality here: we can obviously imagine oaks that stay upright without strong roots, and some oaks keep erect simply by leaning against other trees. But more to the point, peacocks do not *need* gaudy tails in order to survive and reproduce; it is simply that without them their chances of gaining a mate are greatly reduced. Yet surely Foot would want to say that a peacock with a drab tail is defective.

10.4 Conclusion

We make what appear to be perfectly factual judgements about defective traits in plants and animals all the time. What is more, these judgements appear to make implicit reference to some standard of flourishing—what we might think of as the nature of the species. The argument presented in this chapter assumes that the onus is on Foot to explain to us what sorts of things species' natures are, what sorts of facts might make them one way rather than another, how we might decide between two competing judgements about the nature of a given species, and so forth. A biological account of functioning based on fitness seems wholly inappropriate to serve as a basis for claims about flourishing, and Foot's own Aristotelian account is also inadequate. So while it is true to say that we make many confident claims about what makes for organic flourishing, we have yet to be given reasons for thinking that these judgements are of a factual nature. This is not the place to explore the proper understanding of such judgements in detail, but it is worth noting the attractions of a simple error theory (see also Chapter 4). On this view, we mistakenly believe species to have natures of the kind that enable us to make judgements of the form Foot has alerted us to. We might go on to argue that the failure to reduce statements of these kinds to those of a more respectable biological variety does not show the existence of non-natural facts about species natures, nor does it point us in the direction of an Aristotelian account of function. It shows, simply, that we are mistaken about the nature of the natural world. I have not argued in detail for such an error theory here, but I suggest that the problems faced by any account of the alleged facticity of claims about species' flourishing and species' natures boost the plausibility of such accounts considerably.

11
Health, Naturalism, and Policy

11.1 Naturalism about Health

There is an intuitive appeal associated with the idea that the distinction between health and disease is one that matters for ethical purposes. One might think, for example, that medical interventions aimed at treating disease conditions are more worthy, ethically speaking, than interventions that aim to boost a trait beyond what is required for health.[1] Many would accept genetic engineering if used for the treatment of serious diseases, but not (if such a thing were possible) for further augmenting the IQ of a child already within the normal range. Some think that while nationalized health services should provide mandatory treatment for diabetes, cosmetic plastic surgery should rarely be funded out of the public purse. The distinction between health and disease may seem salient here, too. One might think that given a situation of scarce resources to be spent on health, the alteration of non-disease traits should receive much lower priority than the rectification of diseases, no matter how much misery might be relieved by purely cosmetic surgery. Of course, one's extreme misery might itself be pathological, an indication of a mental illness that inflates beyond what is reasonable the importance of appearance. In such cases, treatment might be justified after all, but again this is in virtue of the fact that the individual seeking surgery harbours some form of pathology.

The two broad classes of question mentioned here—one about how far we might be allowed to alter or augment our traits (or the traits of our

---

[1] This chapter appears for the first time here. For comments on various earlier versions I am grateful to Ron Amundson, Christopher Boorse, Marc Ereshefsky, Stephen John, Elselijn Kingma, Karen Neander, and Onora O'Neill.

children), another about the limits to calls on the public purse—are very different. One concerns what should be permitted within a sphere of private action, another what is obligatory on the part of public funding. I will discuss these differences a little towards the end of this chapter; for the moment it is enough to note how the distinction between disease and health, and the correlative distinction between treatment and enhancement, seem to do some ethical work in both cases.

An argument to substantiate the claim that the health/disease distinction is of ethical significance requires two basic steps. First, one must give a generic account of what disease is, in order for us to know how this line is to be drawn. Second, one must show why this line marks a boundary that is ethically salient in itself. So-called naturalistic accounts of health and disease are usually understood as accounts that appeal solely to natural facts—more specifically, to the sorts of facts dealt with by the biological sciences—in drawing this line, and these accounts will be my focus in this chapter.

Before embarking on an investigation of the plausibility of naturalistic accounts of disease, an important preliminary is in order. We should not suppose too quickly that only naturalistic theories can make the health/disease distinction an objective matter. Suppose we claim (rather implausibly) that a disease trait is any physiological condition that severely reduces the well-being of the bearer. This is a normative theory, not a naturalistic one, because well-being is not (on the face of things, at least) a matter of *natural* fact, accounted for solely by reference to concepts from biology. And yet, we might couple this normative theory to a so-called objective list account of well-being, which dissociates an individual's true welfare from their subjective desires or preferences (see e.g. Parfit 1984: appendix I for a valuable taxonomy of different senses of welfare). Such an account of disease would be objective in various senses of the word. It makes the disease status of a trait independent of how it is perceived by its bearer. Assuming that a plausible objective list of welfare determinants can be drawn, it will yield a fairly sharp line between disease and non-disease traits. Even so, this account is normative, since disease is defined in terms of welfare. Normative theories link disease to such concepts as welfare, discomfort, 'badness', or some such. Naturalistic theories instead link disease to biological malfunction (e.g. Boorse 1997; 1977). Yet both normative and naturalistic theories have the potential for objectivity in the sense required for the political task of drawing a relatively sharp line between health and disease.

What I want to consider in this chapter is whether naturalistic theories of disease are able to serve as the basis for views that hold the health/disease distinction to be an ethically salient one in itself. I will argue that they fail in this job. It is true that very many diseases carry titles to treatment, but it is not their disease status itself that confers this title. As a result, bioethicists and political philosophers concerned with health policy should not make use of such naturalistic theories of disease, unless it is to point out the gap between establishing the status of a trait as diseased, and further ethical questions about practical action including obligation to offer treatment, the permissibility of intervention, and so forth. Hence my primary topic is not so much the plausibility of naturalistic theories of disease themselves, but the application of these theories—especially Boorse's theory—in the political arena of health resources allocation (see Daniels 1985; 2008; Buchanan et al. 2000, for examples of such applications). If anything, I expect Boorse to be an ally in this claim of mine. Boorse himself thinks it is a strength of his theory of disease that it explains the divergence between claims about disease and claims about desirability or clinical intervention (1997: 15; 2011: 21). Since clinical intervention is what interests the policy-maker, we should be suspicious of one who uses naturalistic theories of disease as a guide in this matter.

11.2 Boorse on Disease

The most thoroughly articulated version of naturalism about disease is the biostatistical theory of Christopher Boorse. Boorse gives the following analysis of disease (e.g. Boorse 1997: 7–8):

1. The reference class is a natural class of organisms of uniform functional design—specifically, an age group of a sex of a species.
2. A normal function of a part or process within members of the reference class is a statistically typical contribution by it to their individual survival and reproduction.
3. A disease is a type of internal state which is either an impairment of normal functional ability, i.e. a reduction of one or more functional abilities below typical efficiency, or a limitation on functional ability caused by environmental agents.
4. Health is the absence of disease.

The basic idea behind Boorse's view is that disease is an impairment of normal biological function. On Boorse's view, what makes a trait a disease is whether it results in the bearer being unable to perform the usual functions of her reference group, or only being able to perform those functions to a comparatively low level. So if one has a broken leg, or a leg lacking muscles, then this reduces the efficiency with which the usual function of legs—namely, assisting in locomotion—is exercised. Hence one is diseased. It might seem strange to call broken legs or legs that have been damaged since birth 'diseases', rather than injuries and impairments respectively. Boorse is aware of this: he explicitly intends to use the term 'disease' to cover a very wide range of traits, including those more usually thought of as injuries, impairments, and other conditions that amount to a departure from health. In more recent work, Boorse has dropped this broad usage of 'disease' in favour of 'pathology'.

11.3 Pluralistic Naturalism

Before proceeding further, it is instructive to read this passage from Sommerhoff (1950: 6), cited by Boorse as an example of his inspiration in thinking about functions and health:

The beast is not distinguishable from its dung save by the end-serving and integrating activities which unite it into an ordered, self-regulating and single whole, and impart to the individual whole that unique independence from the vicissitudes of the environment and that unique power to hold its own by making internal adjustments, which all living organisms possess in some degree.

Boorse is inspired, then, by a view of organic function as the orchestration of a system so as to give it independence from a varying environment through a range of internal alterations. This raises a question. Why not jump off from this starting point to think of disease not as adverse departure from typical species functioning, but rather as adverse departure from the ability—perhaps the typical overall level of ability—to reproduce and to maintain internal order against environmental alterations? The fact that the wheelchair user is different from the walker, and worse at using her legs for locomotion, does not by itself entail that the wheelchair user is any less able than the walker to discharge Sommerhoff's overall organic goals of internal adjustment and systemic maintenance. In that sense, the wheelchair user need be no worse off, from a biological point of view, than someone who uses her legs to get around. Of course, the person with damaged legs and no wheelchair is reckoned pathological by Sommerhoff's implicit criterion of disease. With no wheelchair, this person cannot easily get from one place to another and is therefore less likely to be able to find nutrition and shelter, and is more vulnerable to changes in the local environment as a result. Even so, use of a wheelchair and access to an appropriate environment (or the construction of an appropriate environment) might put this right. While harbouring an adverse departure from normal function, the wheelchair-bound individual is able to function adequately, albeit abnormally, given the demands of life. Why not endorse a more plural naturalism, according to which statistically subnormal *part* functioning need not be pathological, so long as the *overall* ability of the organism (in combination with technical aids) to survive and reproduce is at a normal level?

It would be a mistake to respond to this more pluralistic conception of health by arguing that when technological aids are required to ensure maintenance and integrity of the organism, the apparent 'health' is only an artificial illusion. That is because even those of us who are clearly healthy rely for our maintenance and integrity on a range of technologies, most obviously such technologies as clothing and shelter. Most of us are unable to survive without clothes, yet we would not say that we all enjoy a fake, artificially sustained health because of this. Nor should we put too much weight on the idea that the elements that sustain an organism should be internal to the organism, or carried with the organism. Humans, as Kim Sterelny frequently points out (e.g. Sterelny 2001; 2003; 2012), are *niche constructors* par excellence. Like other creatures, we ensure our survival not merely by adapting and reacting to an independent environment, but by interacting with, constructing, and modifying that environment so that it becomes one that can sustain us. The building of stairs for leg users is one way of constructing a niche that enables locomotion, the building of wheelchair ramps and wheelchairs is another. Some disability rights activists have argued that their conditions are not properly viewed as pathological at all once properly accommodated (see Amundson 2000; 2005). We can see immediately that it is not only normative, or perhaps social constructivist, conceptions of disease that look likely to facilitate this conclusion.

Another possible response to the suggestion of a more pluralistic naturalism is to point to the distinction between *function* and *functional readiness*. On Boorse's view, it is the possession of species-typical functional readiness that is essential to health (see Kingma 2010 and Hausman 2011 for further discussion). A haemophiliac with no wounds is pathological on Boorse's account, for if that person were to get a cut, she would not be able to respond. Equally, the argument goes, it would be a mistake to argue that individuals with myopia are no longer pathological once they are wearing glasses, or that wheelchair users are no longer pathological once in an environment with ramps. In each case, the requirement of functional readiness is not met. If the myope's glasses were lost or broken she wouldn't be able to see; if ramps couldn't be found, or were broken, the wheelchair user couldn't get around.

This response does not suffice to reject the more pluralistic conception of health. On the view we are considering, we should indeed agree that readiness to withstand some range of alternative environments is important to health. Yet what matters is readiness to withstand environments that the individual in question is likely to encounter. Clearly, it doesn't impair the health of normal individuals if their own respiratory systems could not cope if exposed to an atmosphere containing only nitrogen. Functional unreadiness with respect to that environment does not undo health, because it is not an environment that we have any significant chance of encountering. The haemophiliac is indeed vulnerable in cases where cuts and scratches are likely. Yet the construction of a safe environment that minimizes such probabilities and the reliable availability of doses of clotting factor together constitute a system (one that is technology-based and partly portable) that renders the haemophiliac immune to such risks just as our own internal clotting systems make us immune. Such a view, where technological aids constitute cures for disease, does seem strange, I admit. Yet this awkwardness seems to express nothing more than a bias that makes us call internal, but not external, technological solutions 'cures'. Artificial hips are commonly thought of as cures for those whose natural hips are damaged. The fact that the imagined clotting system is instead carried outside the body doesn't seem good enough grounds to deny it the status of cure, so long as it is indeed as reliable as the more normal system.

So the pluralist analysis allows that individuals equally able to survive and reproduce in the face of likely environmental alterations are equally healthy, even though some may realize this ability in very different ways to others and using external technological aids rather than natural internal organs. The fact that normal individuals may be able to survive in a broader range of environments does not seem relevant. True, leg users can get around on stairs and ramps, wheelchair users only on ramps. Yet what matters to the assessment of overall health in this sense is the ability to survive and reproduce in the face of likely environmental perturbations. Environments that offer widespread access to reliable ramps and stairs will afford equally robust possibilities for locomotion to both wheelchair users and leg users. Environments where these ramps are of poor quality and availability will make wheelchair users less healthy than leg users. The important point to note is that the pluralist naturalistic analysis of health allows that very radical environmental alterations will count as curing disease. Less radical environmental alterations will result in environments in which the locomotive abilities of the wheelchair user are better than before, but still inferior in comparison with those of leg users.

Amundson (2000) gives a number of apparent counterexamples to Boorse that this rather more pluralistic version of naturalism can accommodate. Amundson tries to show that Boorse's account is mistaken by providing biomedically documented examples of individuals who function well, yet in unusual ways. These include a goat that walks perfectly well on its hind legs alone; a man who thinks perfectly well with a brain only a fraction of the usual size; and the large numbers of people who get around perfectly well by using wheelchairs and ramps, instead of legs and stairs. Amundson uses these examples in a set of arguments that try to show three things: first, that there is no such thing as normal function (a concept on which Boorse's theory relies); second, that naturalistic views of disease are flawed; third, that Boorse's theory of disease is flawed. Although Amundson has additional arguments in favour of these conclusions, the examples on their own demonstrate the third claim at most. To show that there is no such thing as normal function is to show that there is no single, prevalent species design. Providing examples of alternative viable designs will not undermine any assertion about the existence of a statistically typical design, even if these examples might show (against Boorse) that one can function abnormally and still be quite healthy. Amundson's examples also leave open a naturalized view of disease that agrees with Boorse in the thought that disease is biological dysfunction, while retaining a more pluralistic conception of the ways in which good overall functioning can be achieved.

Boorse himself will doubtless respond to Amundson, and reject the more pluralistic naturalism I sketched above, by arguing that while one could perhaps construct such a free-wheeling account of pathology, this is not, as a matter of fact, the concept of pathology that underlies the judgements of modern biomedical science. Boorse's aim has always been to understand the implicit disease concept that underlies theoretical pathology; and, he will argue, the very fact that impaired legs *are* viewed as pathological by medical scientists is enough to show that our more liberal account cannot plausibly underlie these modern scientific judgements (Boorse 2011: 20).

11.4 General Functions and Reference Classes

I have been assuming that Boorse's theory tells us that a wheelchair user in an environment with rich provision of ramps is pathological, and I have suggested that this pronouncement fits badly with Boorse's own inspiration from Sommerhoff. In fact, it is unclear exactly what Boorse's theory tells us about the pathological status of the person in the wheelchair. There are two possible ways he could deal with the question, as the following analogous example helps to make clear. Most of us grasp objects with the right hand. This capacity contributes to survival and reproduction by helping us to eat, clean ourselves, and so forth. In virtue of (I suppose) some neural wiring, left-handers are rather bad at manipulating objects with their right hands. Hence neural wiring of left-handers is such as to adversely impair the exercise of a statistically normal function, namely, right-hand manipulation. So does Boorse have to conclude that left-handers are pathological?

Boorse denies that the neural wiring of left-handers is a disease (2011: 22). He might justify this denial by insisting that we should not divide functions too finely. Left-handers' neural wiring impairs *right-hand manipulation*, but not *manipulation*. Left-handers are just as good at manipulating as right-handers. Left-handers are not diseased so long as we take the relevant function to be stated in a suitably general way.

The question for Boorse is whether such 'suitably general' ways can be restricted so as to avoid his own biostatistical theory sliding towards the more pluralistic form of naturalism I explored in the last section. On the face of it, Boorse will say that the person in the wheelchair is diseased because their legs do not contribute to the statistically normal function of walking. However, there is a move that can be made here that looks the same as that which Boorse must take if he is to avoid the conclusion that left-handers are diseased. One might say that the relevant function by which we reckon disease is not *walking* but *locomotion*. The person in the wheelchair is able to discharge this function perfectly well, albeit through a rather different method. It seems that a slippery slope threatens here, at the end of which lies the claim that the only relevant function by which we reckon traits diseased is the wholly general one of maintaining organic integrity against a harsh environment regardless of how, precisely, this is done. Impairments of this general function will count as diseases, but not impairments of many statistically typical modes of carrying out this function, so long as there are compensations for these impairments in other parts of the system. That, after all, is why left-handers are not diseased.

Maybe Boorse can find a way to preserve a middle ground for his theory so that left-handers are ruled out of the disease category but people in wheelchairs are ruled in. He will perhaps argue that so long as left-handers' total hand-and-arm abilities equal those of right-handers, we should think of the two as examples of non-pathological polymorphism (1977: 558). Whether such abilities are equal turns on environmental facts about the availability of scissors, golf clubs, writing desks, and so forth designed for right- and left-handers. This does not remove the problem of how to classify wheelchair users for, by analogous reasoning, it would seem that so long as the overall locomotive abilities of wheelchair users and leg users are equal, we should see these traits as examples of non-pathological polymorphism also. And again, whether such abilities are equal turns on environmental facts about the availability of ramps and so forth. It is a tenet of the disability rights movement that disabilities are classed as such in virtue of features of the social and technological environment in which we live. Amundson's preference for a social constructivist view of disease in service of this claim is unnecessary. Naturalistic theories—perhaps even Boorse's—can support it also.

Another general set of worries for Boorse's theory comes from how we are to individuate 'reference classes'.[2] Obviously, if just any group of individuals can form a reference class, then there will be no diseases. The group of cancer sufferers will constitute a reference class, and suffering from cancer will be statistically normal within that class. Boorse insists that reference classes be 'natural classes of organisms of uniform functional design; specifically, an age group of a sex of a species' (1997: 7). We can press a little on the question of why, for example, the blind do not count as a reference class. The small size of the group cannot be enough to exclude them, for some age groups are also small. Nor, on pain of circularity, can we appeal to the claim that blindness is a disease. We could perhaps exclude the blind on the grounds that they do not differ enough from the sighted in order to qualify as a distinct reference class. The differences between groups perhaps have to be large and diverse; it may not be enough to differ only in one trait. Boorse may have something like this in mind when he writes that 'the reference class was restricted by sex and age because of differences in normal physiology between males and females, young and old' (1997: 8). If this is the response, we must ask whether wheelchair users form a distinct reference class. As I have already suggested, the reliance of an individual on technical aids is no good reason to deny them healthy status, nor is it 'unnatural' to make use of such aids. Since the normal physiology of the wheelchair user is quite different from that of the walker, it doesn't seem so strange to put such individuals into their own reference class. Again, these difficulties in individuating reference classes may push Boorse towards a view of health as the ability to maintain organic integrity in the face of a hostile environment, regardless of how that is achieved and with what technological aids.

More likely, Boorse will deny the requirement to formulate any general account of what appropriate reference classes have in common. Instead, he might simply argue that, as a matter of fact, theoretical pathology *does* demarcate its reference classes by age and sex, but *not* by differences in the ability to see or by differences in physiology with respect to locomotion. Since age and sex can be defined using purely

---

[2] Karen Neander has independently arrived at similar objections (Neander 1984). The problem of reference classes has been presented in detail in an important paper by Elselijn Kingma (2007).

biological language, we should conclude that the pathology concept our scientists have is entirely naturalistic in spite of our ability to conjecture alternative concepts, with alternative reference classes, that an alternative science of pathology might have used. Regardless of which of these responses Boorse employs, we face the thought that such unusual individuals as chimaeric hermaphrodites will not be diseased, for they do not fall into the reference class of either males or females, being too different from both. I will not press these problems any further here (for this, see Kingma 2007), for my primary target is not Boorse's theory of disease, but applications of naturalized theories of disease to questions of health policy.

11.5 Naturalism and Policy

The phenomenon of reproduction exposes problems for attempts to use any naturalistic theory of disease to ground decisions in healthcare ethics. Indeed, some might think that reproduction creates fatal problems for naturalistic accounts of disease themselves. For reasons I will explain in the next section, I will not be arguing here that the equation of disease with adverse departure from biological function cannot be right. My argument will be that if we choose to embrace the consequences to which the naturalistic conception leads regarding what is and what is not a disease, we cannot then plausibly claim that the question of whether a trait is diseased has ethical salience in itself.

Here, in brief, is the argument. Since modern biology is preoccupied with reproduction, it seems that any theory that links disease to biological dysfunction should include the ability to have healthy offspring among an organism's goals and functions, in addition to abilities to maintain its own structure against a hostile environment. The result of this is (probably) to make homosexuality a disease, since homosexuality looks likely to count as a malfunctioning trait on many naturalistic accounts and certainly on Boorse's account (Ruse 1997). Yet the mere fact that homosexuality impairs the ability to reproduce does not tell us anything that would shed light on how we should respond practically to this trait. It does not tell us that it is unpleasant for the bearer, nor that it is harmful to others, nor that it is likely to lead to the death of the bearer, nor that it is likely to impair other aspects of functional performance. What is more, the fact that reproduction is among our biological goals does not entail that reproduction is one of the capacities to which we are all entitled. In brief, the naturalistic disease status of the trait does not tell us whether it is a suitable candidate either for voluntary self-funded treatment or for state-funded treatment. These questions of appropriate action are just what we want to know about when we ask questions of health policy. So the naturalistic status of the trait as diseased is of no relevance in itself to the ethical question of appropriate action. Of course, this brief response ignores an important possibility. Being an Ivy League graduate is neither necessary nor sufficient for being a good employee: there are good employees from other universities and there are Ivy Leaguers who are lazy, dishonest, and so forth. But one might nonetheless argue that being an Ivy League graduate is a decent indicator of being a good employee. Later in this chapter I will consider the analogous question of whether, in spite of the fact that disease status is neither necessary nor sufficient for entitlement to treatment, it might nonetheless be a broadly reliable *indicator* of what is an appropriate target for medical treatment.

The disease status of homosexuality is hard to avoid on all naturalistic theories of disease. On Boorse's view, disease traits are departures from normal function, where functions are contributions to the survival and reproduction of the organism. Hence, on Boorse's view, homosexuality looks likely to be a disease because it is a trait that gives rise to adverse departure from normal contribution to reproduction. This conclusion is not assured, and my argument here is not watertight. It relies, for example, on a conception of homosexuality (or at least some forms of homosexuality) as a distinctive dispositional state. One who thinks there are no homosexuals, only homosexual acts, might resist the claim that homosexuality is a disposition, and hence that it is any kind of state, diseased or otherwise. The argument also relies on the claim that homosexuality impairs the ability of the individual to reproduce. This looks likely to be true—it looks likely that homosexual desires impair chances of the bearer attracting and having sex with people of the opposite sex— yet empirical research could overturn this intuition.[3] Once these additional steps of the argument are filled out, we can see that even such uncommon traits as the inclination to become a Catholic priest may turn

---

[3] Ruse (1997: 155) writes that 'the surveys we have do suggest that [homosexuals] are much more likely to go unmarried, and certainly to parent fewer children'.

out to be diseases on Boorse's view, for here too we have a disposition that reduces the bearer's ability to reproduce below the statistical norm.

Alternative views of disease as dysfunction also have uncomfortable consequences for one who wants the boundary between health and disease to mark an important ethical line. Suppose we think that diseases are malfunctions, where function is understood not in Boorse's terms, but instead in the manner of the now popular 'selected effects' (SE) account of biological function (e.g. Neander 1991). Here, the function of a trait is what it has been selected for. It is, in other words, the effect that earlier instances of the same trait type had that explain the prevalence of the trait today. If some trait fails to have the effect that explains the selection of traits of the same type, then it is said to be malfunctioning.

On this selected-effects view of health, the question of whether homosexuality is a disease will turn on a rather complex set of conceptual and empirical questions. It may be that homosexuality should not be explained in adaptive terms. Certainly at first glance it is hard to see how a trait that (let us assume) makes reproduction unlikely could have spread through a population because of its effects on fitness. If homosexuality does not seem amenable to adaptive explanation, then it is likely that we should understand homosexuality as a trait that fails to have the effect that caused other traits of the same type—forms of heterosexuality or bisexuality—to spread or be maintained in a population. Views that equate disease with SE malfunction may see homosexuality as a pathology.

That conclusion is not guaranteed by the SE account. One may be able to mount an explanation for the maintenance of homosexuality in a population through an argument that uses the inclusive-fitness concept. Perhaps if homosexuals care for the offspring of their siblings, say, then homosexuality can indeed be explained adaptively, although not through effects on the survival and reproduction of the bearers of the trait, but through effects on survival and reproduction of the bearer's close kin (see Vasey et al. 2007 for tentative empirical support and McKnight 1997 for a general review). While the selected-effects account is able to index functions to contributions to inclusive fitness, Boorse's account links functions only to individual prospects of survival and reproduction. That is why homosexuality seems more likely to be reckoned functional on the SE account than on Boorse's account.

Naturalism does not entail that homosexuality is a disease: it makes it an empirical question, to be decided (if we base our naturalism on the SE account of functions) by work in evolutionary biology. Note a couple of awkward consequences that this discussion of the SE view of disease brings out. First, the question of whether a trait is a disease or not will turn on historical investigation. If homosexuality turns out to be a healthy trait, it will probably be because of its past effects on children of earlier homosexuals' relatives. So anyone who wants to make the status of a trait as diseased one that carries normative import is committed to the thought that our obligations to offer treatment for some condition depend not on the current nature of the condition with respect to the individual, but on the history of the trait, including its past effects on individuals other than the bearer. This implies, in turn, that health insurers (for example) should discover whether the treatment of some trait might be subject to compulsory reimbursement by investigating the evolutionary past and assessing mathematical evolutionary models such as those proposed by Gavrilets and Rice (2006).

Second, the shift to an evolutionary view makes pathology hostage to evolutionary enquiry more generally. Suppose detailed empirical work ends up endorsing Thornhill and Palmer's (2000) suggestion that rape is an adaptation. In that case it may also turn out that some low-status males who are not disposed to rape are pathological. Others have suggested that depression—at least mild depression—is an adaptation that reduces impulse to action in cases where it is likely to be dangerous or futile (e.g. Nesse 2000), perhaps making happiness a disease in some circumstances.

Regardless of whether we think the evolutionary view gives the right account of disease, it will make the line between health and disease fall in such a place that it cannot plausibly be thought to have ethical relevance in itself. It threatens to include homosexuality, elevated moods, and propensities to consensual sexual behaviour as diseases, while reckoning depression and a propensity to rape healthy. The case of depression suggests that a quasi-naturalistic theory, according to which diseases are *harmful* dysfunctions, also cannot work to make the health/disease boundary ethically salient (for such an account see Wakefield 1992a; 1992b; 1999). Depression carries a claim for treatment, yet it may not be biologically dysfunctional. (Cooper 2002 makes the same argument.)

11.6 Diagnosis

It is now time to attempt a general diagnosis of why naturalistic theories of disease fail to mark out an ethically salient category. As a first pass, we can say the following. We need a certain range of mental and physical capacities, and we need to be free from debilitating pain, if we are to become what we want to become and achieve what we want to achieve. We might have thought that health is valuable because the state of health overlaps with this set of capacities that enable individual flourishing. What we now see is that naturalistic theories of disease in fact reckon traits healthy according to their ability to promote, not the achievement of individual goals, but reproduction. Healthy traits will include those that promote reproduction at the expense of the survival of the organism, and perhaps those that promote reproduction of relatives at the expense of individual reproduction.

This diagnosis seems to rest on the imperfect overlap of individual goals and biological goals. Even if all organisms have reproduction as a *biological* goal, it is not the case that all people have reproduction in mind as a plan for life or desire. Even if evolutionary reflections persuade us that, in some sense, we aim at maximizing inclusive fitness, we certainly do not all have the reproduction of our kin as a personal psychological goal. Reproduction is important to some of us and not to others, and traits that assist us in finding mates and raising children are also important to some but not to others. Naturalistic theories of health fail to draw an important ethical line because biological goals are only sometimes, not always, the goals of individuals.

The preceding paragraph gets close to Kitcher's (1996: ch. 8) analysis of the failures of naturalistic health definitions for policy purposes. Kitcher points primarily to the divergence of individual goals and desires, and normal biological functions. As he puts it:

Laborious attempts to develop an objective—value free—notion of disease are ultimately of no help, because they ground our understanding of health and disease in facts about our evolutionary history that, although they may sometimes be congruent with our goals for ourselves, are quite external to those goals. (Kitcher 1996: 213–14)

That analysis is on the right lines, but incomplete in this context. It is easier to diagnose why naturalistic theories of health fail to mark an ethically salient boundary in the case of debates about efforts to use genetic engineering to produce 'designer babies'. Here, many commentators have suggested that parents can alter the traits of their children in any way they like so long as they do not harm them (e.g. Harris 1998). Regardless of the truth of this principle, it is clear that the status of some trait as diseased will not, on a naturalistic view, ensure that it is harmful to the individual, and nor will all harmful traits be diseases. The focus of naturalistic theories on reproduction, as well as survival, ensures that many traits that harm the individual can be healthy, so long as they contribute to the reproductive success of the individual in question or the individual's relatives.

A rather different set of questions arises when we ask what medical services should be made available to all. The mismatch between biological goals and individual goals might seem to explain why, for example, psychotherapy should not be made available to all homosexuals. Homosexuals may not care about being able to reproduce, even if reproduction is an important biological goal. Yet this diagnosis fails. Many people do not care about their own survival—they pursue lifestyles that hasten death and degeneration of the body—yet this is not enough to make us think that many medical technologies that sustain survival should not be made available to all. So the fact that a given group of people does not care about some aspect of biologically defined health is not enough to show that medical services that will restore health should not be made available to all, should the individual choose it.

One of the reasons that we think it awkward to raise the possibility of the obligatory provision of funding to restore homosexuals to proper reproductive health is that we assume such intervention would be one of psychotherapeutic or psychopharmacological cure. In fact, we might also solve the problem of the homosexual's reduced ability to reproduce by giving that individual access to a range of technologies, or social services, that enable this person to raise a child, perhaps one with some close genetic relatedness. The question of whether this intervention counts as curing the homosexual's disease will turn on those awkward questions I raised in Section 11.3 regarding whether our naturalism should reckon health a state of statistically typical function, or instead one that enables survival and reproduction regardless of the mode of functioning and the technological aids employed. While Boorse's theory probably continues to call the homosexual with access to reproductive technologies diseased, the more pluralistic, technologically permissive naturalism sketched earlier will more likely consider this individual reproductively healthy.

It is at least possible, then, for the naturalist to make the following kind of argument: we should strive as far as is possible to make available to all individuals a state of health. That is to say, all individuals should have resources made available to them that enable their efficient survival and reproduction. This does not entail the alteration of the mind of the homosexual, only the provision of a range of technological resources that will enable the homosexual to have a child should he or she decide to do so. It does not seem far-fetched to imagine the gay community itself backing this argument, as a way of securing access to reproductive technologies for same-sex couples.

Such a suggestion looks to make the naturalistic distinction between disease and health an ethically salient one once more. It certainly takes us a long way from a crude and indefensible view that begins with Boorse's view of disease and suggests (a suggestion that Boorse himself emphatically rejects) that all and only Boorsean diseases have a title to medical treatment. I can only offer a few short comments in response to it. The assertion that all should be entitled to the capacity to survive and reproduce, because these are naturalistically healthy states, relies on a fairly crude sociobiological fallacy. It is an open question whether, and in what sense, reproduction is a right, and that question cannot be answered by appealing to the fact that reproduction is a biological goal. Reproduction involves bringing another individual into the world. Because of this it is a serious business, more than simply a facet of our 'self-expression', since self-expression concerns how we choose to order only our own lives. While we may rightly condemn coercive attempts to stop people from reproducing, this does not entail that we have any duty to assist all individuals in creating a new person.[4]

11.7 Types of Disease

The conclusion that naturalistic concepts of disease cannot make the health/disease distinction one of ethical salience needs to be tempered a little. A certain subset of naturalistic diseases—the degenerative

---

[4] See O'Neill (2002) for an elaboration and defence of these arguments.

diseases—have an important and wide-ranging impact on our ability to achieve any of our goals. They lead to the erosion and disintegration of many bodily functions to the point where they can make all action impossible and ultimately hasten death. What is more, in many cases it is clearly medical treatment that these conditions warrant, for there is no possibility of the condition being accommodated by environmental modification. Conditions that cause severe pain are similar: they, too, can render a life unliveable, and medical treatment may be the only means to the improvement of the lives of sufferers.

Disabilities often differ from painful and degenerative diseases in both respects—the disabled can often be effectively accommodated, rather than treated, and the lives of the disabled can be as rich as those of the functionally normal. Norman Daniels (1985; 2008) has tried to give a rationale for our concern with all types of disease through the suggestion that very many diseases, understood naturalistically, limit opportunity. That suggestion seems quite likely, and furthermore, limitations of opportunity do carry ethical import. However, that does not show that the disease status of a trait itself is what carries ethical import. Many traits that are not Boorsean naturalistic disease traits limit opportunity (unwanted pregnancies, perhaps mild depression), and many traits that are Boorsean naturalistic diseases do not (or should not) limit opportunity. Homosexuality is just such a trait. Homosexuality may limit opportunities to reproduce, even if it makes no difference to the bearer's abilities to access favoured social and economic positions; but we have seen that it is questionable whether we have any entitlement to reproduction, and in any case this is hardly the kind of opportunity that those who champion equality of opportunity have typically been keen to defend. Daniels's theory of healthcare resource allocation is concerned with ensuring equal opportunities for all through the provision of medical care and, more recently, through just arrangements of all the various determinants of health. Daniels is aware that in grounding the value of health in the concern for equality of opportunity he must acknowledge that diseases are not always bad, for they do not always reduce opportunity, and that many non-disease traits should be avoided, for they do reduce opportunity. The fact of whether, on a naturalistic theory, some trait is diseased or not is irrelevant to issues about equal opportunity that motivate Daniels's view.

11.8 Disease and Treatment

Why don't I argue that these problem cases for applying naturalistic theories of disease to questions of health policy show that there is something wrong with naturalistic accounts of disease themselves? Surely if Boorse's account, for example, tells us that homosexuality is a pathology, then this example shows us that the theory is flawed. Doesn't Boorse's account just give the wrong answer about homosexuality? (This is the conclusion that Cooper 2002 reaches using many of the same examples.)

Boorse has a response to this line of criticism that reinforces my conclusion about the inappropriate status of naturalistic concepts of disease for informing health policy. Boorse is adamant that the disease concept is value-free (see Radick 2001 for stress on this important theme). Boorse's assertion sits very uneasily with any attempt to use the disease concept in healthcare ethics, unless it is to free the disease concept from carrying important ethical weight. Boorse in fact recognizes that homosexuality is a disease on his account (at least he hints at this recognition quite strongly):

It can hardly be denied that one normal function of sexual desires is to promote reproduction. (Boorse 1975: 63)

But:

The desirability of having species-typical desires is not nearly so obvious as the desirability of having species-typical physiological functions. (Boorse 1975: 63)

In other words, the association between being a disease and being undesirable is imperfect and contingent. Homosexuality is a Boorsean disease in virtue of being a departure from normal functioning, yet it might be a preferable trait for some all the same. Good health, on Boorse's view, is not always desirable. Programmes that make oral contraception widely available are also, on Boorse's view, programmes that encourage disease, but diseases that are desirable and liberating nonetheless.

So Boorse has an answer to the claim that homosexuality is a counterexample to his theory. He will tell us that we wrongly think it cannot be a disease because we correctly think there is nothing wrong with it. That is, we think it is a trait that often causes no problems for its bearers, one that many may value, and that they would rather not lose. We therefore think it is a trait that, in many cases, should not be altered. However, on Boorse's thoroughly value-free theory of disease, it is simply a mistake to think that because some trait is often desirable, and not an appropriate target for medical correction, it cannot be a disease. By severing the link between disease and treatment, Boorse himself makes clear the dubious relevance of his naturalistic disease concept to the questions of medical action that concern bioethicists and health policy-makers.

In an early piece Boorse tried to forge a link not between *disease* and treatment, but between *illness* and treatment. (He has since revised his understanding of illness.) If this move works, then the ethical salience of naturalistic concepts of health might be partially reinstated, since illnesses are defined in terms of disease. Boorse defined illnesses as those diseases that are incapacitating. He held that illness, unlike disease, is a normative concept:

A disease is an *illness* only if it is serious enough to be incapacitating, and therefore is:

(i) undesirable for its bearer;
(ii) a title to special treatment; and
(iii) a valid excuse for normally criticizable behavior (Boorse 1975: 61).

Not all diseases are incapacitating or undesirable: homosexuality is a Boorsean disease, but not always incapacitating, so often not an illness. Note that I say 'not always': for those homosexuals who are frustrated in their desires to have children their sexuality is, presumably, incapacitating and hence a Boorsean illness among those who want a genetically related family (see Ruse 1997: 160–1).

It is true, for reasons we will shortly see, that having species-typical physiological functions is often desirable for the individual in question. Departure from typical functioning is therefore likely to frustrate the desires of many. The result is that many Boorsean diseases are incapacitating and many aspects of health are desirable. Even so, Boorse's condition (ii) does not follow from the fact that some trait is incapacitating (his use of 'therefore' suggests he once believed it did), unless 'special treatment' is read in a very wide way to point to a need for some intervention or another to alleviate incapacity, rather than medical intervention. We can see why this is so by constructing an engine for the generation of problem cases for Boorse's theory.

Boorsean diseases can be generated through the following recipe. First, take a trait that is statistically unusual. Second, imagine a world where environments are such that this trait generally leads to impaired survival or reproduction by comparison with the more normal trait. The trait will be diseased in that world on Boorse's account. Now, since Boorse's account makes no mention of the causal pathways by which these environments make one trait lead to impaired functioning by comparison with the other, traits that impair functioning owing to social conventions geared up for typical traits will count as diseased. So far as I can see, the skin colour of black people is a disease trait on Boorse's view, in hypothetical cases where commonly encountered social institutions make the attainment of wealth, and hence partners and healthy children, a difficulty. In such worlds, skin colour leads to an adverse departure from normal functioning.

We can now appreciate that one reason that normal functioning is often desirable for the individual is that social environments and institutions may be set up to cater for the physiology and psychology of the majority. Those with different physiologies and psychologies will find life difficult in comparison with that majority. Boorse's view of disease, by severing the link between disease and suitability for treatment, says nothing about how social and medical programmes should respond to the existence of individuals who depart adversely from normal functioning. We have a choice as to whether we try to alter society in such a way that the functional majority no longer enjoys special advantages, or whether we alter individuals to bring their makeup closer to that of the majority. Since skin colour in a racist world is incapacitating for black people (in the sense that it prevents them from realizing many of their goals), it seems that we should also say that their skin colour is a Boorsean illness. Yet it is clear in this case, at least, that if any 'special treatment' is demanded, it is treatment of the institutions that make the lives of black people difficult, not treatment of skin colour. Boorse did not say enough about what he meant by 'special treatment' to let us know whether he thought the existence of illness calls for some intervention or another, or whether it specifically calls for medical treatment of the ill individual. I will not try to speculate about where he stood on this question.

The proper answer to the question of how to intervene in the face of incapacity turns on complex ethical considerations that are not themselves informed by thinking about whether a trait is a disease or an illness. Instead, we will have to look to general economic and welfarist concerns: is it more efficient to keep environments constant and treat individuals, or should we effect an alteration of the environment instead? We also need to look to questions of rights and duties. Some have argued that we have no general duty to ensure that all can reproduce; yet the failure of reproductive functioning may be incapacitating all the same. So some disease traits may hamper the exercise of desired normal functions that no one has a duty to maintain. The race case also makes clear the importance of avoiding a society in which traits are diseases (or illnesses) in virtue of discriminatory treatment of the bearers of those traits. These arguments show that the question of whether some trait is a (naturalistic) disease or illness has no relevance in itself to determining the answer to questions such as whether some trait should be subject to medical cure through state funding. The claims that some group suffers through social injustice and discriminatory treatment at the hands of others, and that their suffering should be relieved by social, not medical, intervention, are compatible with the claim that their suffering is brought about by a naturalistic disease.

11.9 Daniels on Healthcare

I mentioned near the beginning of this chapter that Norman Daniels has made use of Boorse's theory of health in his important discussion of health, healthcare, and justice. The endorsement of Boorse was quite explicit in his *Just Health Care* (Daniels 1985). His later work, *Just Health*, is rather more neutral, claiming that 'Boorse's strict naturalism is not important to my argument' (2008: 39), but Daniels's attraction to that naturalism remains evident. Because Boorse's specific view is not discussed in so much detail in Daniels's 2008 work, I will refer primarily to the more thorough discussion in his 1985 book in what follows.

Daniels's overall theory of health and justice is remarkably sophisticated. What is more, in spite of the fact that Daniels answers the question 'Is health of special moral importance?' with a resounding 'Yes!', on my reading Daniels does *not* argue that departures from health are *always* of moral concern, or even that health per se is of special moral importance.

Instead, Daniels aims to explain why departures from health are so often of moral concern. The answer he gives is that they often restrict opportunities in an unjust manner. But he is not committed to the thought that departures from health are always of moral concern, nor to the thought that the type of moral importance we accord to the alleviation of disease only applies to the alleviation of disease. Daniels aims to explain, in other words, why we are rightly concerned to relieve many diseases, without arguing that the health/disease distinction coincides with an important moral boundary.

Note, as a corollary of this, that Daniels does not simply take Boorse's theory of health at face value, before proceeding to argue that since disease traits reduce opportunity, all Boorsean diseases carry an urgent claim to rectification via medical cure. Daniels notices that Boorse's theory threatens to make skin colour a disease in a racist world. He proposes that we must modify Boorse's theory by specifying 'the range of environments taken as "natural" for the purpose of revealing dysfunction [...] If we allow too much of the social environment, then racially discriminatory environments might make being of the wrong race a disease' (1985: 30). This specification of normal environments cannot save Daniels from the conclusion that homosexuality is a disease, so long as he endorses Boorse's basic view. That may not bother him, for he might try to argue in the manner of Section 11.4 that homosexuals are entitled to equal reproductive opportunities compared with heterosexuals, where this is achieved through the provision of reproductive technologies. This concern aside, the most obvious way for Daniels to specify which environments lead to genuine diseases is to exclude from the causal generation of disease those social environments that are ethically objectionable. Daniels tries to dodge the problem of specifying appropriate environments, but the most likely way for him to reach his desired endpoint is to make the disease concept normative, by asserting that disease traits are those traits that result in impairment of normal function in non-discriminatory environments.

After declining to comment on just how we should revise Boorse's account of disease to yield appropriately normative conclusions about intervention, Daniels falls back on an appeal to accepted practices in drawing the line between disease and health:

These difficult issues need not detain us. My discussion does not turn on this deeper, strong claim about non-normativeness advanced by some advocates of the biomedical model. It is enough for my purposes that the line between disease and the absence of disease is, for the general run of cases, uncontroversial and ascertainable through publicly acceptable methods, such as those of the biomedical sciences. (1985: 30)

Now here the disability rights activist should be worried. The biomedical sciences will certainly find a physiological difference between wheelchair users and walkers, and these sciences may also find that these differences correlate strongly with opportunity loss. What is more, it may be the view of the general public that wheelchair users are indeed deficient in some way. So, many of Daniels's criteria for disease-hood seem to be satisfied. Even so, consider again the case of race in a prejudiced society. Here, too, the biomedical sciences will find a physiological difference between blacks and whites in terms of skin colour, and one that correlates with opportunity loss. If the society is strongly prejudiced, it is also possible that the general public will see blacks as diseased or deficient. Meeting Daniels's tests here does not answer the crucial question for health policy of whether discrimination plays a role in lowering the opportunity of some group of physiologically different individuals.

I do not claim in this chapter to defend the arguments of disability rights groups. What the chapter does show is the ground on which these arguments need to be played out. The crucial question at stake in arguments over whether deaf parents should be allowed to deafen their children, or whether funds should be used to screen out paraplegia using genetic tests or instead to accommodate paraplegics in the built environment, is not the question of whether deafness and paraplegia are diseases. The significant questions for the resolution of these debates instead include those of whether the maintenance of a built environment to favour one group over another that is physiologically different, or the maintenance of a set of communicative practices that disadvantage those without hearing, are truly objectionable forms of discrimination. In the race case we are likely to be on the side of the black parent who says that although her child will be disadvantaged compared with a white child, she will refuse interventions to alter that child's skin colour, lobbying instead for social reform. In some circumstances, refusing intervention that could benefit the child is justified. The question for deaf parents who wish to refuse treatment that could prevent their child from becoming deaf, in favour of lobbying for social reform that will give the deaf equal opportunities, is whether the two cases are relevantly similar in terms of how social practices bias against the minority. It is of no help to either side in this debate to argue over the claim that deafness is a disease but skin colour is not. That proposition is irrelevant.

11.10 Healthcare Without Health

In denying the direct relevance of a naturalized concept of health to questions of health policy, I am suggesting that policy-makers should instead look to considerations of the distribution of welfare, or opportunity, or resources in determining the allocation of medical and social goods. Here my argument endorses the conclusion of Ereshefsky (2009), albeit via a very different route. Exactly how policy-makers should reason about the allocation of freedoms and resources will depend on complex arguments about the currency of distributive justice—arguments I have neither the space nor the expertise to discuss here—but we should attempt a preliminary exploration of the terrain. It might appear that, in giving up a naturalized theory of health as the ground of decision-making in health policy, 'anything goes', and that the intuitions that seem to underpin the commonly believed ethical salience of the treatment/enhancement distinction must go also. Daniels seems to share this worry when he says: 'the normative approach I do consider a threat to public agreement is not constrained by an independent account of departures from normal functioning' (2008: 40). Perhaps he is concerned that once we jettison appeals to naturalized disease, we will have to admit that if my average-shaped nose makes me miserable, then considerations of welfare distribution give me equal entitlement to treatment over someone whose cancer makes them equally miserable. If my slightly lower than average height restricts opportunities, then I have as strong a claim to an operation to extend my spine as a coronary patient does to a heart bypass.

In fact, these conclusions would all be too hasty. We have already seen how the distinction between treating disease and altering the environment to reduce the incapacity that results from disease might enable one to stick to the thought that spine extensions for those with just-below-average height should not be funded while cancer care should. If those with shorter than average height are penalized because employers, say, have a tendency to undervalue their skills and knowledge, then it is better to divert scarce resources to removing this prejudice than to extending spinal length. What is more, if we suspect that social values are such that taller members of society are treated better than shorter members, then so long as this situation continues, those who are shorter will always be at an opportunity loss no matter how tall they might be in absolute terms. General economic considerations will then recommend that the cheapest long-run way to equalize opportunity is to remove its root cause through social intervention instead of constantly pushing up mean height through medical intervention. Looking directly at the political considerations that might restrict certain interventions is illuminating, because as well as showing that not just anything goes, we also move far beyond a simplistic position that asserts that diseases demand medical intervention from public funds, while enhancements should be funded from personal sources. The general economic arguments, and considerations of eliminating prejudice, that tell against spinal extension will also push us to seeking environmental and social measures, instead of medical intervention, to combat some disabilities.

The case of the nose operation is harder to resolve. The problem of 'champagne preferences' is commonly taken to show that those with expensive tastes are not owed greater resources than those with more modest tastes (e.g. Scanlon 1975; Dworkin 2000). We should not try to equalize welfare between a lover of sailing and a lover of skittles, the argument goes, by giving the sailor the considerable resources necessary to purchase a yacht while leaving the skittles player content with the tiny funds needed to cover his match entry fee. Reasoning along similar lines can exempt from mandatory funding the cosmetic operation that an individual wants from the 'champagne preference' for a perfectly formed nose. There is no salient difference, from the perspective of justice, between the overwhelming desire for the perfect nose and the overwhelming desire for the perfect yacht. If a theory of justice has machinery to enable us to withhold the resources necessary for the satisfaction of the latter preference, then it can also withhold resources necessary for the satisfaction of the former.

Doesn't this show that the coronary patient who wants an expensive working heart also has no such entitlement? Isn't this desire a champagne preference too? Here we might reinstate an asymmetry by looking at the full consequences of the operations to be considered. In one case, denying an individual the perfect nose they have always wanted will affect only some of their plans for life. Yet if an individual is denied a working heart, there is no possibility of them successfully carrying out any life plans at all.

Our nose sufferer might respond by telling us that the situations are symmetrical after all. He cannot imagine living any kind of life unless it is with a perfect nose; without it, all of his plans will fail as surely as all the plans of the individual with the blocked arteries will fail unless an operation is approved. Even here there are resources to drive a wedge between the two cases that do not look to the disease status of the conditions to justify our attitudes towards them. First, if the nose sufferer's misery truly is as comprehensive as he claims, we might think our money more prudently spent persuading him that appearance is less important than he thinks, rather than risk further expensive demands for operations to the ears, lips, and so forth. Here a concern for welfare reinstates an asymmetry between operating on a heart and operating on a nose, even if it keeps open the view that the nose sufferer is owed psychological therapy.

Other ways of preserving an asymmetry between the operations include pointing to an objective account of welfare that allows us to rule out the preference for the perfect nose as one that is too frivolous to have any urgency in its satisfaction (see Scanlon 1975), although obviously the ground of this conception of objective welfare must not lie in naturalized health if the conclusions of this chapter are to stand. One might also point to responsibility to ground the asymmetry. Perhaps, one could argue, the preference for a perfect nose is one that the agent is responsible for. The agent could, reasonably, have had other, cheaper preferences. On the other hand, it is not a matter of choice that a functioning heart (as opposed to some cheaper alternative) helps in one's life plans. The heart is a 'forced move' for any self-determining agent. Obviously I am not offering anything like a well-worked-through set of answers to any of these problems here. I want only to hint at some of the ways in which various schools of ethical thought might discriminate between, say, cosmetic surgery and other forms of surgery without relying on a prior naturalized distinction between enhancement and treatment.

Styles of ethical reasoning that stress the value of autonomy can also restrict the scope of mandatory healthcare to the maintenance of those traits that are necessary for an individual to formulate and realize a range of reasonable life plans. On such a view, we would expect all to be given basic physical and psychological capacities, since these capacities are necessary to the formulation and achievement of all reasonable life plans. However, resource constraints might then limit the apportionment of medical care for the maintenance and provision of traits beyond what is needed for such generic functioning.

This suggestion merits further attention. It might seem that it returns us to something like Boorse's own conception of health. Aren't these 'generic functionings' just the same as those traits that enable the survival and reproduction of the organism? Won't traits that enable 'generic functioning' simply overlap with Boorse's own healthy traits? While there may be some similarities in extension of the concepts, their motivations are quite different. 'Generic functionings' are those traits that enable all reasonable life plans. Many naturalized disease traits are unlikely to have much impact, if any, on such plans. Athlete's foot does not destroy our autonomy and neither does homosexuality. Note also that although a concern with generic biological function is almost certain to include the ability to reproduce among some set of basic capacities, we have already seen that the more political question of what capacities all should be entitled to problematizes the 'right to reproduce' in useful ways.

Putting value on autonomy might also explain in a rather different way the adherence of some to a treatment/enhancement distinction in the case of decisions over 'designer babies'. Those who value autonomy might see no harm in parents equipping their children with genes that enable the kinds of capacities suitable to all reasonable life plans the child is likely to want to make, while opposing parental interventions that exert undue influence over the precise life plans the child is likely to take on. The reasons for adhering to a rough version of the treatment/ enhancement distinction in the genetic engineering case are very different from the reasons that might lead us to support a rough treatment/ enhancement distinction in the case of surgical treatment of adults. It is hard to see how augmenting traits like attention or memory well beyond the norm could undermine the autonomy of a child. On the other hand, it seems unlikely that very high levels of attention or memory would form part of a decent minimum of capacities that we would make available to all through state-funded medical care. What is permitted as a parental intervention, and what is prioritized as a matter of state-funded care, will be quite different given this concern with autonomy. Focusing on the political motivations that underlie attitudes towards healthcare enables much richer debate over medical interventions than simply pointing to the distinction between health and disease, and its supposed ethical import.

References

Alcock, J. (2001). *The Triumph of Sociobiology* (Oxford: Oxford University Press).

Amundson, R. (2000). 'Against Normal Function', *Studies in History and Philosophy of Biological and Biomedical Sciences* 31: 33–53.

Amundson, R. (2005). 'Disability, Ideology, and Quality of Life: A Bias in Biomedical Ethics', in D. Wasserman, J. Bickenbach, and R. Wachbroit (eds), *Quality of Life and Human Difference: Genetic Testing, Health Care and Disability* (Cambridge: Cambridge University Press), 101–24.

Anderson, J., Johnstone, B., and Remley, D. (1999). 'Breast-Feeding and Cognitive Development: A Meta-analysis', *American Journal of Clinical Nutrition* 70: 525–35.

Andrianantoandro, E., Basu, S., Karig, D., and Weiss, R. (2006). 'Synthetic Biology: New Engineering Rules for an Emerging Discipline', *Molecular Systems Biology* 2, doi: 10.1038/msb4100073.

Ariew, A. (1996). 'Innateness and Canalization', *Philosophy of Science* 63: S19–S27.

Ariew, A. (1999). 'Innateness Is Canalization: A Defence of a Developmental Account of Innateness', in V. Hardcastle (ed.), *Biology Meets Psychology: Conjectures, Connections, Constraints* (Cambridge, Mass.: MIT Press), 117–38.

Arkin, A., and Fletcher, D. (2006). 'Fast, Cheap, and Somewhat in Control', *Genome Biology* 7: 114.

Arneson, R. (1989). 'Equality and Equal Opportunity for Welfare', *Philosophical Studies* 56: 77–93.

Ashcroft, R. (2003). 'American Biofutures: Ideology and Utopia in the Fukuyama/Stock Debate', *Journal of Medical Ethics* 29: 59–62.

Atran, S. (1990). *Cognitive Foundations of Natural History* (Cambridge: Cambridge University Press).

Atran, S., Estin, P., Coley, J., and Medin, D. (1997). 'Generic Species and Basic Levels: Essence and Appearance in Folk Biology', *Journal of Ethnobiology* 17: 17–43.

Barker, M. (2010). 'Specious Intrinsicalism', *Philosophy of Science* 77: 73–91.

Basalla, G. (1988). *The Evolution of Technology* (Cambridge: Cambridge University Press).

Bateson, P. (1987). 'Biological Approaches to the Study of Development', *International Journal of Behavioral Development* 10: 1–22.

Bateson, P., and Martin, P. (1999). *Design for a Life: How Behaviour Develops* (London: Cape).

Bateson, W. (1904). 'Practical Aspects of the New Discoveries in Heredity', in *Proceedings: International Conference on Plant Breeding and Hybridization* (New York: Horticultural Society of New York), 1–10.

Blackburn, S. (1998). *Ruling Passions: A Theory of Practical Reasoning* (Oxford: Oxford University Press).

BMA (2007). *Boosting Your Brainpower: Ethical Aspects of Cognitive Enhancements* (London: British Medical Association).

Boldt, J., and Müller, O. (2008). 'Newtons of the Leaves of Grass', *Nature Biotechnology* 26: 387–9.

Boorse, C. (1975). 'On the Distinction Between Disease and Illness', *Philosophy and Public Affairs* 5: 49–68.

Boorse, C. (1977). 'Health as a Theoretical Concept', *Philosophy of Science* 44: 542–73.

Boorse, C. (1997). 'A Rebuttal on Health', in J. M. Humber and R. F. Almeder (eds), *What Is Disease?* (Totowa, NJ: Humana Press), 3–134.

Boorse, C. (2011). 'Concepts of Health and Disease', in F. Gifford (ed.), *Handbook of the Philosophy of Science*, vol. 16: *Philosophy of Medicine* (Oxford: Elsevier), 13–64.

Bostrom, N. (2003). 'Human Genetic Enhancements: A Transhumanist Perspective', *Journal of Value Inquiry* 37: 493–506.

Bredenoord, A., Dondorp, W., Pennings, G., and de Wert, G. (2011). 'Ethics of Modifying the Mitochondrial Genome', *Journal of Medical Ethics* 37: 97–100.

Briggle, A. (2010). *A Rich Bioethics: Public Policy, Biotechnology and the Kass Council* (Notre Dame, Ind.: University of Notre Dame Press).

Buchanan, A. (2009). 'Human Nature and Enhancement', *Bioethics* 23: 141–50.

Buchanan, A., Brock, D., Daniels, N., and Wikler, D. (2000). *From Chance to Choice: Genetics and Justice* (Cambridge: Cambridge University Press).

Buller, D. (1999). 'DeFreuding Evolutionary Psychology: Adaptation and Human Motivation', in V. Hardcastle (ed.), *Where Biology Meets Psychology: Philosophical Essays* (Cambridge, Mass.: MIT Press), 99–114.

Buss, D. (1999). *Evolutionary Psychology: The New Science of the Mind* (Boston: Allyn and Bacon).

Camacho, D., and Collins, J. (2009). 'Systems Biology Strikes Gold', *Cell* 137: 24–6.

Caspi, A., et al. (2002). 'Role of Genotype in the Cycle of Violence in Maltreated Children', *Science* 297: 851–4.

Caspi, A., et al. (2007). 'Moderation of Breastfeeding Effects on the IQ by Genetic Variation in Fatty Acid Metabolism', *PNAS* 104: 18860–5.

Clark, A., and Chalmers, D. (1998). 'The Extended Mind', *Analysis* 58: 10–23.

Cobb, R., Sun, N., and Zhao, H. (2012). 'Directed Evolution and a Powerful Synthetic Biology Tool', *Methods*, http://dx.doi.org/10.1016/j.ymeth.2012.03.009.

Cohen, G. (1989). 'On the Currency of Egalitarian Justice', *Ethics* 99: 914–23.

Cohen, G. (1993). 'Equality of What? On Welfare, Goods and Capabilities', in M. Nussbaum (ed.), *The Quality of Life* (Oxford: Clarendon Press), 54–61.

Cooper, R. (2002). 'Disease', *Studies in History and Philosophy of Biological and Biomedical Sciences* 33: 263–82.

Cosmides, L., and Tooby, J. (1997). Letter, *New York Review of Books*, 7 July.

Cosmides, L., Tooby, J., and Barkow, J. (1992). 'Introduction: Evolutionary Psychology and Conceptual Integration', in J. Barkow, L. Cosmides, and J. Tooby (eds), *The Adapted Mind* (Oxford: Oxford University Press).

Cronin, H., and Curry, O. (2000a). 'The Evolved Family', in H. Wilkinson (ed.), *Family Business, Demos Collection 15* (London: Demos), 151–7.

Cronin, H., and Curry, O. (2000b). 'Pity Poor Men', *Guardian*, Saturday 5 February.

Cross, N. (2000). *Engineering Design Methods: Strategies for Product Design*, 3rd edn (Chichester: Wiley).

Curry, O., Cronin, H., and Ashworth, J. (eds) (1996). 'Matters of Life and Death: The Worldview from Evolutionary Psychology', *Demos Quarterly 10*.

Daly, M., and Wilson, M. (1988). *Homicide* (New York: De Gruyter).

Daly, M., and Wilson, M. (1998). *The Truth About Cinderella* (London: Weidenfeld and Nicolson).

Daniels, N. (1985). *Just Health Care* (Cambridge: Cambridge University Press).

Daniels, N. (2008). *Just Health* (Cambridge: Cambridge University Press).

Daniels, N. (2009). 'Can Anyone Really Be Talking About Ethically Modifying Human Nature?', in J. Savulescu and N. Bostrom (eds), *Human Enhancement* (Oxford: Oxford University Press), 25–42.

Daniels, N., Kennedy, B., and Kawachi, I. (2000). 'Justice is Good for Our Health', *Boston Review* 25: 4–19.

Darwin, C. (1859). *The Origin of Species* (London: John Murray).

Darwin, C. (1868). *The Variation of Animals and Plants under Domestication* (London: John Murray).

Darwin, C. (1877/2004). *The Descent of Man*, 2nd edn, edited and introduced by J. Moore and A. Desmond (London: Penguin).

Davies, P. S. (1999). 'The Conflict of Evolutionary Psychology', in V. Hardcastle (ed.), *Where Biology Meets Philosophy: Philosophical Essays* (Cambridge, Mass.: MIT Press), 67–81.

Dawkins, R. (1982). *The Extended Phenotype* (Oxford: Oxford University Press).

Department of Health (2014). 'Mitochondrial Donation: A Consultation on Draft Regulations to Permit the Use of New Treatment Techniques to Prevent the Transmission of a Serious Mitochondrial Disease from Mother to Child', UK

Department of Health, 27 February https://www.gov.uk/government/consultations/serious-mitochondrial-disease-new-techniques-to-prevent-transmission.

Devitt, M. (2008). 'Resurrecting Biological Essentialism', *Philosophy of Science* 75: 344–82.

Devitt, M. (2010). 'Species Have (Partly) Intrinsic Essences', *Philosophy of Science* 77: 648–61.

Dougherty, M., and Arnold, F. (2009). 'Directed Evolution: New Parts and Optimized Function', *Current Opinion in Biotechnology* 20: 486–91.

Dupré, J. (1981). 'Natural Kinds and Biological Taxa', *Philosophical Review* 90: 66–91.

Dupré, J. (2012). *Processes of Life* (Oxford: Oxford University Press).

Dworkin, R. (1981). 'What is Equality? Part Two: Equality of Resources', *Philosophy and Public Affairs* 10: 283–345.

Dworkin, R. (2000). *Sovereign Virtue* (Cambridge, Mass.: Harvard University Press).

Endy, D. (2005). 'Foundations for Engineering Biology', *Nature* 438: 449–53.

Ereshefsky, M. (2001). *The Poverty of the Linnean Hierarchy* (Cambridge: Cambridge University Press).

Ereshefsky, M. (2008). 'Systematics and Taxonomy', in S. Sarkar and A. Plutynski (eds), *The Blackwell Companion to the Philosophy of Biology* (Oxford: Blackwell), 99–118.

Ereshefsky, M. (2009). 'Defining "Health" and "Disease"', *Studies in History and Philosophy of Biological and Biomedical Sciences* 40: 221–7.

Ereshefsky, M. (2010). 'What's Wrong with the New Biological Essentialism?', *Philosophy of Science* 77: 674–85.

Ereshefsky, M., and Matthen, M. (2005). 'Taxonomy, Polymorphism, and History: An Introduction to Population Structure Theory', *Philosophy of Science* 72: 1–21.

European Commission (2005). *Synthetic Biology: Applying Engineering to Biology. Report of a NEST High-Level Expert Group* (Brussels: European Commission), ftp://ftp.cordis.europa.eu/pub/nest/docs/syntheticbiology_b5_eur21796_en.pdf.

Faris, S. (2010). 'Breeding Ancient Cattle Back from Extinction', *Time*, 12 February.

Fodor, J. (2004). 'Water's Water Everywhere', *London Review of Books* 26(20): 17–19.

Foot, P. (1961). 'Goodness and Choice', *Proceedings of the Aristotelian Society*, Supplementary Volume 35: 45–60.

Foot, P. (2001). *Natural Goodness* (Oxford: Oxford University Press).

Forbes, G. (1986). 'In Defense of Absolute Essentialism', *Midwest Studies in Philosophy* 11: 3–31.

Gavrilets, S., and Rice, W. (2006). 'Genetic Models of Homosexuality: Generating Testable Predictions', *Proceedings of the Royal Society B* 273: 3031–8.

Gelman, S., and Hirschfeld, L. (1999). 'How Biological Is Essentialism?', in S. Atran and D. Medin (eds), *Folkbiology* (Cambridge, Mass.: MIT Press), 403–45.

Ghiselin, M. (1997). *Metaphysics and the Origin of Species* (Albany, NY: SUNY Press).

Godfrey-Smith, P. (2000a). 'On the Theoretical Role of "Genetic Coding"', *Philosophy of Science* 67: 26–44.

Godfrey-Smith, P. (2000b). 'Information, Arbitrariness, and Selection: Comments on Maynard Smith', *Philosophy of Science* 67: 202–7.

Godfrey-Smith, P. (2007). 'Information in Biology', in D. Hull and M. Ruse (eds), *The Cambridge Companion to the Philosophy of Biology* (Cambridge: Cambridge University Press), 103–19.

Gray, R. (1992). 'Death of the Gene: Developmental Systems Strike Back', in P. Griffiths (ed.), *Trees of Life: Essays in Philosophy of Biology* (Dordrecht: Kluwer), 165–209.

Griffiths, P. (1999). 'Squaring the Circle: Natural Kinds with Historical Essences', in R. Wilson (ed.), *Species: New Interdisciplinary Essays* (Cambridge, Mass.: MIT Press), 209–28.

Griffiths, P. (2002). 'What is Innateness?', *Monist* 85: 70–85.

Griffiths, P., and Gray, R. (1994). 'Developmental Systems and Evolutionary Explanation', *Journal of Philosophy* 91: 277–304.

Griffiths, P., and Gray, R. (1997). 'Replicator II: Judgment Day', *Biology and Philosophy* 12: 471–92.

Habermas, J. (2003). *The Future of Human Nature* (Cambridge: Polity Press).

Harris, J. (1998). 'Rights and Reproductive Choice', in S. Holm and J. Harris (eds), *The Future of Human Reproduction* (Oxford: Oxford University Press), 5–37.

Harris, J. (2007). *Enhancing Evolution: The Ethical Case for Making Better People* (Princeton, NJ: Princeton University Press).

Harris, J., and Holm, S. (2002). 'Extending Human Lifespan and the Precautionary Paradox', *Journal of Medicine and Philosophy* 27: 355–68.

Hausman, D. (2011). 'Is an Overdose of Paracetamol Bad for One's Health?', *British Journal for the Philosophy of Science* 62: 656–68.

Heck, H. (1951). 'The Breeding-Back of the Aurochs', *Oryx* 1: 117–22.

Heyes, C. (2012). 'What's Social About Social Learning?', *Journal of Comparative Psychology* 126: 193–202.

Holm, S. (1999). 'There Is Nothing Special about Genetic Information', in L. Thompson and R. Chadwick (eds), *Genetic Information* (New York: Kluwer), 97–104.

Hope, T., and McMillan, J. (2003). 'Ethical Problems before Conception', *The Lancet* 361: 2164.

Houkes, W., and Vermaas, P. (2010). *Technical Functions: On the Use and Design of Artefacts* (Berlin: Springer).

House of Commons General Committee Debates 3 June 2008. Human Fertilisation and Embryology Bill [Lords] Column 22. (Online) UK Parliament. http://www.publications.parliament.uk/pa/cm200708/cmpublic/human/080603/am/80603s01.htm#end. Accessed 4 September 2013.

House of Lords Hansard Debates 3 December 2007. Human Fertilisation and Embryology Bill Column 1506. (Online) UK Parliament. http://www.publications.parliament.uk/pa/ld200708/ldhansrd/text/71203-0004.htm. Accessed 4 September 2013.

HSE (2007). *HSE Horizon Scanning Intelligence Group Short Report: Synthetic Biology* (London: Health and Safety Executive). http://www.hse.gov.uk/horizons/assets/documents/synthetic.pdf.

Hughes, K., and Reynolds, R. (2005). 'Evolutionary and Mechanistic Theories of Aging', *Annual Review of Entomology* 50: 421–45.

Hull, D. (1986). 'Human Nature', *PSA: Proceedings of the Biennial Meeting of the Philosophy of Science Association* 2: 3–13.

John, S. (2007). 'How to Take Deontological Concerns Seriously in Risk–Cost–Benefit Analysis: A Reinterpretation of the Precautionary Principle', *Journal of Medical Ethics* 33: 221–4.

Kass, L. (1998). 'The Wisdom of Repugnance: Why We Should Ban the Cloning of Humans', *Valparaiso University Law Review* 32: 679–705.

Kass, L. (2003). 'Ageless Bodies, Happy Souls: Biotechnology and the Pursuit of Perfection', *New Atlantis* 1: 9–28.

Kauffman, S. (1996). *At Home in the Universe* (London: Penguin).

Kevles, D. (1985). *In the Name of Eugenics* (Berkeley: University of California Press).

Kingma, E. (2007). 'What Is It to Be Healthy?', *Analysis* 67: 128–33.

Kingma, E. (2010). 'Paracetamol, Poison, and Polio: Why Boorse's Account of Function Fails to Distinguish Health and Disease', *British Journal for the Philosophy of Science* 61: 241–64.

Kitcher, P. (1985). *Vaulting Ambition* (Cambridge, Mass.: MIT Press).

Kitcher, P. (1996). *The Lives to Come: The Genetic Revolution and Human Possibilities* (London: Allen Lane).

Kripke, S. (1980). *Naming and Necessity* (Oxford: Blackwell).

Kwok, R. (2010). 'Five Hard Truths for Synthetic Biology', *Nature* 463: 288–90.

Laland, K., Odling-Smee, J., and Feldman, M. (2000). 'Niche Construction, Biological Evolution, and Cultural Change', *Behavioral and Brain Sciences* 23: 131–75.

Levins, R., and Lewontin, R. (1985). *The Dialectical Biologist* (Cambridge, Mass.: Harvard University Press).

Lewens, T. (2002). 'Adaptationism and Engineering', *Biology and Philosophy* 17: 1–31.
Lewens, T. (2004a). *Organisms and Artifacts: Design in Nature and Elsewhere* (Cambridge, Mass.: MIT Press).
Lewens, T. (2004b). 'What Is Genethics?', *Journal of Medical Ethics* 30: 326–8.
Lewens, T. (2007a). 'Functions', in M. Matthen and C. Stephens (eds), *Handbook of Philosophy of Biology* (New York: Elsevier), 537–59.
Lewens, T. (2007b). *Darwin* (London: Routledge).
Lewens, T. (2008). 'Taking Sensible Precautions', *The Lancet* 371 (9629): 1992–3.
Lewens, T. (2009). 'What Is Wrong with Typological Thinking?', *Philosophy of Science* 76: 355–71.
Lewens, T. (2012). 'Species, Essence, and Explanation', *Studies in History and Philosophy of Biological and Biomedical Sciences* 43: 751–7.
Lewontin, R. (1974). 'The Analysis of Variance and the Analysis of Causes', *American Journal of Human Genetics* 26: 400–11.
Linquist, S., Machery, E., Griffiths, P., and Stotz, K. (2011). 'Exploring the Folkbiological Conception of Human Nature', *Philosophical Transactions of the Royal Society B* 366: 444–53.
Lippert-Rasmussen, K. (2004). 'Are Some Inequalities More Unequal than Others? Nature, Nurture, and Equality', *Utilitas* 16: 193–219.
Lynch, M. (2007). 'The Frailty of Adaptive Hypotheses for the Origins of Organismal Complexity', *PNAS* 104: 8597–604.
MacCormack, A., Rusnak, J., and Baldwin, C. (2008). 'Exploring the Duality between Product and Organizational Architectures: A Test of the "Mirroring" Hypothesis', *Harvard Business School Working Paper*, http://hbswk.hbs.edu/item/5894.html.
Machery, E. (2008). 'A Plea for Human Nature', *Philosophical Psychology* 21: 321–9.
Mameli, M., and Bateson, P. (2006). 'Innateness and the Sciences', *Biology and Philosophy* 21: 155–88.
Manson, N. A. (2002). 'Formulating the Precautionary Principle', *Environmental Ethics* 24: 263–4.
Maynard Smith, J. (2000). 'The Concept of Information in Biology', *Philosophy of Science* 67: 177–94.
McCarthy, D. (2001). 'Why Sex Selection Should Be Legal', *Journal of Medical Ethics* 27: 302–7.
McKnight, J. (1997). *Straight Science? Homosexuality, Evolution, and Adaptation* (London: Routledge).
Mellor, D. H. (1995). *The Facts of Causation* (London: Routledge).
Millet, K. (2011). 'Caesura, Continuity, and Myth: The Stakes of Tethering the Holocaust to German Colonial Theory', in V. Langbehn and M. Salama (eds),

German Colonialism: Race, The Holocaust, and Postwar Germany (New York: Columbia University Press), 146–63.

Millikan, R. (1984). *Language, Thought and Other Biological Categories* (Cambridge, Mass.: MIT Press).

Morange, M. (2009). 'Synthetic Biology: A Bridge between Functional and Evolutionary Biology', *Biological Theory* 4: 368–77.

Müller-Wille, S. (2012). 'Revisiting the Mendelian Revolution', presentation, 19 January, Dept of History and Philosophy of Science, University of Cambridge.

Murray, T. (1997). 'Genetic Exceptionalism and Future Diaries: Is Genetic Information Different from Other Forms of Medical Information?', in M. Rothstein (ed.), *Genetic Secrets: Protecting Privacy and Confidentiality* (New Haven, Conn.: Yale University Press), 60–76.

Nagel, T. (1997). 'Justice and Nature', *Oxford Journal of Legal Studies* 17: 303–21.

Neander, K. (1984). 'Abnormal Psychobiology', Ph.D. dissertation, LaTrobe University.

Neander, K. (1991). 'The Teleological Notion of "Function"', *Australasian Journal of Philosophy* 69: 454–68.

Nesse, R. (2000). 'Is Depression an Adaptation?', *Arch. Gen. Psychiatry* 57: 14–20.

Nicholson, N. (2000). *Managing the Human Animal* (Oakland, Calif.: Texere).

Nuffield Council on Bioethics (2012). *Novel Techniques for the Prevention of Mitochondrial DNA Disorders: An Ethical Review* (London: Nuffield Council on Bioethics).

Odling-Smee, J., Laland, K., and Feldman, M. (1996). 'Niche Construction', *American Naturalist* 147: 641–8.

Okasha, S. (2002). 'Darwinian Metaphysics: Species and the Question of Essentialism', *Synthese* 131: 191–213.

O'Malley, M. (2011). 'Exploration, Iterativity, and Kludging in Synthetic Biology', *Comptes rendus chimie* 14: 406–12.

O'Malley, M., Powell, A., Davies, J., and Calvert, C. (2007). 'Knowledge-Making Distinctions in Synthetic Biology', *BioEssays* 30: 57–65.

O'Neill, O. (2002). *Autonomy and Trust in Bioethics* (Cambridge: Cambridge University Press).

Oyama, S. (2000a). *The Ontogeny of Information* (Durham, NC: Duke University Press).

Oyama, S. (2000b). *Evolution's Eye: A Systems View of the Biology–Culture Divide* (Durham, NC: Duke University Press).

Oyama, S., Griffiths, P., and Gray, R. (eds) (2001). *Cycles of Contingency: Developmental Systems and Evolution* (Cambridge, Mass.: MIT Press).

Papineau, D. (1993). *Philosophical Naturalism* (Oxford: Blackwell).

Parfit, D. (1984). *Reasons and Persons* (Oxford: Oxford University Press).

Paul, D. (1998). 'PKU Screening: Competing Agendas, Diverging Stories', in *The Politics of Heredity* (Albany, NY: SUNY Press), 173–86.

Paul, D. (1999). 'What Is a Genetic Test and Why Does It Matter?', *Endeavour* 23: 159–61.

Pavlicev, M., Cheverud, J., and Wagner, G. (2011). 'Evolution of Adaptive Phenotypic Variation Patterns by Direct Selection for Evolvability', *Proceedings of the Royal Society B* 278: 1903–12.

Peterson, M. (2006). 'The Precautionary Principle Is Incoherent', *Risk Analysis* 26: 595–601.

Peterson, M. (2007). 'Should the Precautionary Principle Guide Our Actions or Our Beliefs?', *Journal of Medical Ethics* 33: 5–10.

Pogge, T. (1989). *Realizing Rawls* (Ithaca, NY: Cornell University Press).

Pogge, T. (2004). 'Relational Conceptions of Justice: Responsibilities for Health Outcomes', in S. Anand, F. Peter, and A. Sen (eds), *Public Health, Ethics, and Equity* (Oxford: Clarendon Press), 135–61.

Porcar, M. (2010). 'Beyond Directed Evolution: Darwinian Selection and a Tool for Synthetic Biology', *Systems and Synthetic Biology* 4: 16.

POST (2008). *Synthetic Biology* (Parliamentary Office of Science and Technology. London: HM Government), http://www.parliament.uk/documents/post/postpn298.pdf.

Radcliffe-Richards, J. (2000). *Human Nature after Darwin* (London: Routledge).

Radick, G. (2001). 'A Critique of Kitcher on Eugenic Reasoning', *Studies in History and Philosophy of Biological and Biomedical Sciences* 32: 741–51.

Randall, D., Schneerson, J., Plaha, K., and File, S. (2003). 'Modafinil Affects Mood, but Not Cognitive Function, in Healthy Young Volunteers', *Human Psychopharmacology* 18: 163–73.

Rawls, J. (1971). *A Theory of Justice* (Oxford: Oxford University Press).

Rawls, J. (1993). *Political Liberalism* (New York: Columbia University Press).

Rawls, J. (1999). *A Theory of Justice*, revised edn (Cambridge, Mass.: Harvard University Press).

Ray, E., and Heyes, C. (2011). 'Imitation in Infancy: The Wealth of the Stimulus', *Developmental Science* 14: 92–105.

Richards, M. P. M. (2001). 'How Distinctive Is Genetic Information?', *Studies in History and Philosophy of Biological and Biomedical Sciences* 32: 663–88.

Ridley, M. (1999). *Genome: The Autobiography of a Species in 23 Chapters* (London: Fourth Estate).

Robertson, T. (1998). 'Possibilities and the Argument for Origin Essentialism', *Mind* 107: 729–49.

Roemer, J. (1996). *Theories of Distributive Justice* (Cambridge, Mass.: Harvard University Press).

Rosenberg, A. (2000). 'The Political Philosophy of Biological Endowments', in *Darwinism in Philosophy, Social Science, and Policy* (Cambridge: Cambridge University Press), 195–225.

Ruse, M. (1997). 'Defining Disease: The Question of Sexual Orientation', in J. M. Humber and R. F. Almeder (eds), *What is Disease?* (Totowa, NJ: Humana Press), 137–71.

Russell, N. (1986). *Like Engend'ring Like: Heredity and Animal Breeding in Early Modern England* (Cambridge: Cambridge University Press).

Salmon, N. (1981). *Reference and Essence* (Princeton, NJ: Princeton University Press).

Sandel, M. (2007). *The Case Against Perfection: Ethics in the Age of Genetic Engineering* (Cambridge, Mass.: Harvard University Press).

Sandin, P. (2007). 'Common-Sense Precaution and Varieties of the Precautionary Principle', in T. Lewens (ed.), *Risk: Philosophical Perspectives* (London: Routledge).

Sandin, P., Peterson, M., Hansson, S. O., Rudén, C., and Juthe, A. (2002). 'Five Charges Against the Precautionary Principle', *Journal of Risk Research* 5: 287–99.

Savulescu, J., and Bostrom, N. (eds) (2009). *Human Enhancement* (Oxford: Oxford University Press).

Scanlon, T. (1975). 'Preference and Urgency', *Journal of Philosophy* 72: 655–69.

Scanlon, T. (1989). 'A Good Start: Reply to Roemer', *Boston Review* 20: 8–9.

Segerstråle, U. (2000). *Defenders of the Truth: The Sociobiology Debate* (Oxford: Oxford University Press).

Sen, A. (1992). *Inequality Reexamined* (Cambridge, Mass.: Harvard University Press).

Shea, N. (2007). 'Representation in the Genome and in Other Inheritance Systems', *Biology and Philosophy* 22: 313–31.

Shea, N. (2012). 'New Thinking, Innateness, and Inherited Representation', *Philosophical Transactions of the Royal Society B* 367: 2234–44.

Shea, N. (2013). 'Inherited Representations Are Read in Development', *British Journal for the Philosophy of Science* 64: 1–31.

Shuster, S. (1987). 'Alternative Reproductive Behaviors: Three Discrete Male Morphs in *Paracerceis Sculpta*', *Journal of Crustacean Biology* 7: 318–27.

Singer, P. (1999). *A Darwinian Left: Politics, Evolution and Cooperation* (London: Weidenfeld and Nicolson).

Sober, E. (1980). 'Evolution, Population Thinking, and Essentialism', *Philosophy of Science* 47: 350–83.

Sober, E. (1988). 'Apportioning Causal Responsibility', *Journal of Philosophy* 85: 303–18.

Sober, E. (2000). 'The Meaning of Genetic Causation', appendix 1 of A. Buchanan, D. Brock, N. Daniels, and D. Wikler, *From Chance to Choice: Genetics and Justice* (Cambridge: Cambridge University Press).

Sober, E. (2001). 'Separating Nature and Nurture', in D. Wasserman and R. Wachbroit (eds), *Genetics and Criminal Behavior: Methods, Meanings and Morals* (Cambridge: Cambridge University Press), 47–78.

Sommerhoff, G. (1950). *Analytical Biology* (Oxford: Oxford University Press).

Stepan, N. (1991). *'The Hour of Eugenics': Race, Gender, and Nation in Latin America* (Ithaca, NY: Cornell University Press).

Sterelny, K. (2001). 'Niche Construction, Developmental Systems, and the Extended Replicator', in S. Oyama, P. Griffiths, and R. Gray (eds), *Cycles of Contingency: Developmental Systems and Evolution* (Cambridge, Mass.: MIT Press), 333–49.

Sterelny, K. (2003). *Thought in a Hostile World* (Oxford: Blackwell).

Sterelny, K. (2012). *The Evolved Apprentice* (Cambridge, Mass.: MIT Press).

Sterelny, K., and Griffiths, P. (1999). *Sex and Death: An Introduction to the Philosophy of Biology* (Chicago: University of Chicago Press).

Sterelny, K., and Kitcher, P. (1988). 'The Return of the Gene', *Journal of Philosophy* 85: 339–60.

Sterelny, K., Smith, K., and Dickison, M. (1996). 'The Extended Replicator', *Biology and Philosophy* 11: 377–403.

Stirling, A. (2003). 'Risk, Uncertainty, and Precaution: Some Instrumental Implications from the Social Sciences', in F. Berkhout, M. Leach, and I. Scoones (eds), *Negotiating Environmental Change: New Perspectives from Social Science* (Cheltenham: Edward Elgar).

Stirling, A. (2005). 'Opening Up or Closing Down: Analysis, Participation, and Power in the Social Appraisal of Technology', in M. Leach, I. Scoones, and B. Wynne (eds), *Science, Citizenship, and Globalisation* (London: Zed), 218–31.

Sunstein, C. (2005). *Laws of Fear: Beyond the Precautionary Principle* (Cambridge: Cambridge University Press).

Temme, K., Zhao, D., and Voigt, C. (2012). 'Refactoring the Nitrogen Fixation Gene Cluster from *Klebsiella oyxtoca*', *PNAS* 109: 7085–90.

Thompson, A. (1997). 'An Evolved Circuit, Intrinsic in Silicon, Entwined with Physics', in T. Higuchi et al. (eds), *Evolvable Systems: From Biology to Hardware* (Berlin: Springer), 390–405.

Thompson, M. (1995). 'The Representation of Life', in R. Hursthouse, G. Lawrence, and W. Quinn (eds), *Virtues and Reasons* (Oxford: Oxford University Press), 247–97.

Thompson, M. (2008). *Life and Action: Elementary Structures of Practice and Practical Thought* (Cambridge, Mass.: Harvard University Press).

Thornhill, R., and Palmer, C. (2000). *A Natural History of Rape: Biological Bases of Sexual Coercion* (Cambridge, Mass.: MIT Press).

Tooby, J., and Cosmides, L. (1992). 'The Psychological Foundations of Culture', in J. Barkow, L. Cosmides, and J. Tooby (eds), *The Adapted Mind* (Oxford: Oxford University Press), 19–136.

Turner, D., Robbins, T., Clarke, L., Aron, A., Dowson, J., and Sahakian, B. (2003). 'Cognitive Enhancing Effects of Modafinil in Healthy Volunteers', *Psychopharmacology* 165: 260–9.

Turner, J. S. (2000). *The Extended Organism* (Cambridge, Mass.: Harvard University Press).

UNESCO (1997). *Universal Declaration on the Human Genome and Human Rights*, http://portal.unesco.org/en/ev.php-URL_ID=13177&URL_DO=DO_TOPIC&URL_SECTION=201.html. Accessed 4 September 2013.

Van Vuure, C. (2005). *Retracing the Aurochs: History, Morphology, and Ecology of an Extinct Wild Ox* (Sofia: Pensoft).

Vasey, P., Pocock, D., and VanderLaan, D. (2007). 'Kin Selection and Male Androphilia in Samoan Fa'afafine', *Evolution and Human Behavior* 28: 159–67.

Vincenti, W. (1990). *What Engineers Know and How They Know It: Analytical Studies from Aeronautical History* (Baltimore, Md.: Johns Hopkins University Press).

Waddington, C. H. (1957). *The Strategy of the Genes: A Discussion of Some Aspects of Theoretical Biology* (London: Ruskin House/George Allen and Unwin).

Wakefield, J. (1992a). 'The Concept of Mental Disorder: On the Boundary between Biological Facts and Social Value', *American Psychologist* 47: 373–88.

Wakefield, J. (1992b). 'Disorder as Harmful Dysfunction: A Conceptual Critique of DSM-III-R's Definition of Mental Disorder', *Psychological Review* 99: 232–47.

Wakefield, J. (1999). 'Evolutionary versus Prototype Analyses of the Concept of Disorder', *Journal of Abnormal Psychology* 108: 374–99.

Wasserman, D., and Wachbroit, R. (eds) (2001). *Genetics and Criminal Behavior: Methods, Meanings, and Morals* (Cambridge: Cambridge University Press).

Winther, R. (2001). 'August Weismann on Germ-Plasm Variation', *Journal of the History of Biology* 34: 517–55.

Ziman, J. (2000). *Technological Innovation as an Evolutionary Process* (Cambridge: Cambridge University Press).

Index

abnormality 179, 181
activism, disability rights, *see* disability
adaptation 10, 36, 46, 68, 74, 124–30, 132, 136, 138–43, 145, 171, 173, 188
adoption (of children) 58, 136
aggression 32, 129–30, 151
algorithms, genetic 71
allelomorphs 65
altruism 138
amino acids 21, 88–9, 107
Amundson, Ron 117–18, 179, 181–3
ANOVA (Analysis of Variance) 146, 154–9
anthropology 46, 52
Aristotelianism 2, 8–9, 11, 40, 50–5, 59, 169–70, 172–4
artefacts 36, 52–3, 62–3, 69, 72–5
artificial selection 62–3, 66–70; *see also* natural selection
asexual reproduction, *see* reproduction
athletic ability 26, 33, 44, 160, 165
Atran, Scott 52
aurochs 66–7
autism 118
autonomy 97, 202–3

Bakewell, Robert 63, 70
Bateson, William 65–6
beavers 128, 145
Beckham, David 96
bestiality 140
Biobricks 60, 68–9
biomarkers 12
biosecurity 78
biostatistical theory of disease 2, 177–8, 180–7, 190–7, 202
bisexuality 187
Blackburn, Simon 3–4
blindness 146–7, 184
Boorse, Christopher 2, 165–78, 180–7, 190–7, 202
Bortolotti, Lisa 2
Boyd, Robert 18, 46
breast-feeding 144, 150–1, 160

Bredenoord, Annelien 89–90
bricolage 68
Briggle, Adam 67
British Medical Association (BMA) 31–2
Buchanan, Allen 3, 55, 58, 146–9, 153–4, 156–67
Buller, David 140, 141

canalization 98, 127, 164
cancer 104, 120, 184, 199
Caspi, Avshalom 151
Catholicism 49, 186
cattle 63
  Heck cattle 66–7
champagne preferences 200
chemistry, synthetic 65
child abuse 134–8, 151
children 20–4, 25–9, 136–8, 141–2, 188–90, 194–5
  adopted 136
  childhood 151
  grandchildren 5–6
  stepchildren 134–5, 137–8
  see also offspring
chimaeric hermaphrodites 185
chromosomes 4–5, 7, 100, 115–16
  X-chromosome 114–16
  Y-chromosome 115
clinical trials 33–4
cloning 56–68
code, genetic 88–9, 106–7
cognition 31, 48, 89, 131–2, 139–40
  cognitive abilities 31, 57, 117
  cognitive adaptations 142
  cognitive biases 55
  cognitive development 21, 52, 133, 160
conception 18, 79, 83–6, 92, 111–12
consent, informed 7
constraint, developmental 46, 56, 132
contraception 193
cooperation 195
corn 154, 157–8

cosmetic surgery 175, 200–1
Cosmides, Leda 127, 131–3
creativity 67–9
criminality 121
Cronin, Helena 125–6
cuckoldry 138
cultural inheritance, *see* inheritance
Curry, Oliver 125–6
cystic fibrosis 22
cytoplasm 4

Daly, Martin 132, 134–9
Danaus plexippus (monarch butterfly) 55
Daniels, Norman 1, 2, 3, 34, 44, 149, 192, 196–9
Darwin, Charles 17–18, 54, 67–8, 70, 142
Dawkins, Richard 114
deafness 89–90, 198–9
death 98–9, 171, 173, 185, 190, 192
Demos 124
depletion 91
depression 132, 188, 192
designer babies 22, 116, 190, 202
determinism, genetic 57, 71, 101, 111, 136
determinism, social 3
developmental biology 12, 142
developmental constraint, *see* constraint, developmental
developmental genetics 131
developmental information 88, 107–8
developmental psychology 131
developmental systems theory 106
Devitt, Michael 42–3
diabetes 175
diet 21–2, 100–1, 108, 116, 118, 161
 calorie-restricted 31
 dietary supplements 25–6
disability 22, 24, 79, 179, 183, 192, 198, 200
 disability rights activism 97, 119, 179, 183, 198
discrimination 118, 121, 129, 148–9, 196–8
disease 2–3, 7, 11, 19, 21–4, 79, 105, 116–17, 121, 132, 161, 166, 175–203
 congenital 3, 21
 degenerative 191–2
distributive justice, *see* justice

divorce 135
DNA 5–6, 58, 61, 76, 86, 98–100, 106, 112, 121
domestication 62, 67
drift, genetic 41, 44, 46
Dupré, John 1, 73
duty 109–10, 150, 191, 196
Dworkin, Ronald 145
dysfunction, *see* function
dyslexia 145

education 7, 18–19, 26, 49, 116, 122–3, 145–6, 152–7, 166–8
egalitarianism 110
embryo selection:
 for deafness 90
 for sex 79
emotivism 4
enhancement 3, 8–9, 11–13, 17–38, 40, 44, 55, 58–9, 61, 77, 176, 199–202
epigenetics 7
epistasis 99
equality:
 of opportunity 97, 109–10, 117–19, 148–50, 153, 163, 167, 192, 197–200
 of resources 96–7, 109–19, 122–3, 199–200
 of welfare 109–10, 119, 199
 see also inequality
Ereshefsky, Marc 199
essentialism 9, 40–3, 52–5, 59, 79–87, 89–92, 111–12, 129
 micro-essences 42
 teleo-essences 52–3
ethnobiology 52, 54
eugenics 18–19, 98, 122
evil 2, 11, 169–70
evolution 1–2, 10, 18, 35–6, 39, 43–50, 62, 67–70, 73–4, 88, 106–8, 124–43, 171–3, 188–9
evolutionary electronics 9, 62, 70–7
evolutionary information 88
evolutionary psychology 10, 43, 45–6, 124–43, 188–9
expressivism 4

feminism 128
fictionalism 11
field guides 44–5, 54–5
fitness 35–6, 72, 132–5, 170–1, 173–4, 187, 189

INDEX 219

fitness landscapes 35–6
inclusive fitness 187, 189
flourishing 3, 11, 21, 55, 58, 118, 170–2, 174, 189
Fodor, Jerry 79
folic acid 18
folk biology 52–5
folk psychology 40, 135–6, 144
Foot, Philippa 2, 10–11, 51–3, 169–74
football 48, 95–6
function:
 Aristotelian 169–74
 biological 11, 34–6, 53, 72–3, 75–7, 140, 145, 172–3, 176–90, 192–7, 199, 201–2
 cognitive 32, 105
 dysfunction 182, 185, 187–8, 194–7, 199
 malfunction 11, 138, 171, 176, 185, 187
functional analysis 132–4, 140
functional genomics 121
functional mode 117–19
functional readiness 180

gametes 4, 9, 31, 65, 80–5, 87, 89, 91–2, 109, 111–12
Gattaca 120–1
gender 56
genes:
 BRCA1 104
 FADS2 151
 Huntington's gene 105
 mitochondrial genes 3–6
 PKU genes 21–2
 role in identity 9–10, 41–3, 80–3, 86–7, 89–90, 92, 111–12
genetic algorithms 71
genetic blueprints 107, 115
genetic engineering 6, 10, 13, 49, 68, 90, 96–7, 109–10, 113–19, 122, 175, 190, 202
genetic enhancement 12, 17–19, 25, 29–30
genetic exceptionalism 12, 96–8, 109, 116, 119–22
genetic information 81, 87–8, 106–8, 113, 120–1
genetic inheritance, *see* inheritance
genetic luck 145

genetic modification 3, 5–7, 21, 33, 61, 68, 70, 76, 78, 89–90, 106, 109, 116, 118, 130
genetic mutation 31, 35, 41, 101, 112
genetic predisposition 151
genetic tests 12, 198
genomes 4–6, 41, 61, 67–8, 70, 87–9, 97, 100, 106, 115, 120–1
 mitochondrial genome 4, 7, 89
genotype 35, 42, 87, 101–3, 115, 117, 150–2, 154–5, 158, 167
germ cell 5, 7, 65
germ-line 5–7, 65, 89–90, 110
germ plasm 6
Ghiselin, Michael 39–42
giftedness 8, 20–1, 56
God:
 children as gifts from God 20
 playing God 27, 76–7, 113
Godfrey-Smith, Peter 88–9, 107

haemophilia 180
handicap 145–7
Harris, Evan 5–6
Harris, John 19, 21, 26–38
healthcare 149, 185, 192–3, 196, 199, 203
Heck cattle 67
Heck, Heinz 66–7
Heck, Lutz 66–7
heredity 66, 106
heterosexuality 187, 197
Heyes, Cecilia 48
homicide, *see* murder
homosexuality 95, 185–94, 197, 202
House of Lords 5
Hull, David 39–40, 42–4, 59
Human Fertilisation and Embryology Bill 5
human nature 2, 8–9, 11, 17–20, 24, 26–7, 39–51, 53, 55–9, 125–7, 129–31, 144–5, 163

identity 12, 79, 81, 84, 86–92, 111
 non-identity problem 9, 79–81, 90–2
illness 21, 194–6
 mental illness 175
 see also depression
imitation 48, 88
inclusive fitness, *see* fitness

inequality 3, 10–11, 37–8, 129–30, 144–68
  natural inequality 3, 10–11, 145–54, 156–68
  social inequality 3, 10–11, 145–52, 154, 156–7, 159–65, 167–8
  see also equality
infanticide, see murder
information:
  personal 12, 121
  see also developmental information; genetic information
inheritance:
  cultural inheritance 18, 69, 113–14, 128
  genetic inheritance 5–6, 18, 113–14
  inheritance systems 88
innateness 10, 52, 95, 98, 105–6, 130, 144, 163–4
insurance 120–1, 188
interactionism 10, 12, 106, 122

Jacob, François 68
jealousy 138
justice 10, 36, 96–7, 106, 109–14, 116, 119, 122–3, 145–50, 153, 157, 160–5, 167–8, 196–7, 199–200

Kass, Leon 2, 17, 26, 55–8
Kauffman, Stuart 36
kinds 40–2, 52
Kingma, Elselijn 184
Kitcher, Philip 100, 135, 189
kludges 77–8
knitting 114–16
Kripke, Saul 42, 79–80, 82, 86, 111–12

lactose tolerance 62
Levins, Richard 65
Lewontin, Richard 65, 155
Lippert-Rasmussen, Kasper 154–9
locomotion 178–9, 181, 183–4
lotteries 113, 145, 147–9, 152, 165
love:
  parental 20–3, 136–8, 141
  romantic 58, 137

Machery, Edouard 40, 43–50
malfunction, see function

Mameli, Matteo 2
Marx, Karl 50, 129
maternal spindle transfer 4, 86
mental Illness, see illness, depression
mitochondria 4–7, 86
  see also genomes
modafinil 31–3
modularity 66, 68, 72–7
  mental modules 140–2
monoamine oxidase A (MAOA) 151
mood 31–2, 188
murder 142
  homicide 125, 134
  infanticide 141–2
myopia 180

Nagel, Thomas 147–8, 164–8
narcolepsy 31
naturalism (about disease) 175–83, 185–6, 188–94, 196, 199, 201–2
natural selection 18, 41, 43–5, 67–71, 74–5, 88, 107–8, 127–8, 131–2, 187;
  see also artificial selection
natural/social distinction 3, 10–11, 56–7, 144–8, 150–4, 156–68
Neander, Karen 184
niche construction 128–9, 179
niches (ecological) 42
nomological conception of human nature 43–4, 49
non-identity problem, see identity
norm of reaction 101–5, 117, 148, 152–3, 160
nucleotides 88, 107

oak trees 169–70, 173
offspring:
  genetically related 45, 136–7, 190, 194
  see also children; parents
Okasha, Samir 40–2
O'Malley, Maureen 61, 69–70, 77–8
O'Neill, Onora 191
ontogeny 95, 144
oocytes 86
oppression 129
organelles 4

paraplegia 119, 198
parental investment 134, 136–8

parents 20–4, 28–30, 58, 79–83, 85–92, 111–12, 132, 134–8, 190
  adoptive parents 58, 136
  foster parents 136
  procrustean parenting 8, 23–4
  step-parents 134–5, 137–8
Parfit, Derek 90–2, 111, 176
parity principle 96–8, 106, 109, 119, 122–3
parthenogenesis 58
paternalism 8, 27, 30
Peterson, Martin 34
phenylalanine 21–2
phenylketonuria (PKU) 21, 161
philandering 45, 124
Pogge, Thomas 146–7, 151, 166
polymorphism 40, 45–6, 183
Powell, Russell 2
precautionary principle 7–8, 25, 27, 29, 32–3, 35–6
pregnancy 141, 192
prejudice 129–30, 198, 200
privacy 121
progeny testing 63
promiscuity 126
Provigil, *see* modafinil

race 54, 118, 195–8
Radick, Gregory 193
rape 125, 139–43, 188
rational design 9, 60–2, 64–75, 77–8
Rawls, John 3, 110, 145–8, 165–6
reproduction 11, 56, 90–1, 135, 139–41, 170–4, 185–7, 189–92, 202
  asexual reproduction 57–8
  sexual reproduction 56–9
reproductive technologies 2, 89–91, 190–1, 197
Richerson, Peter 18, 46
Ridley, Matt 114
risk 8, 25–30, 32, 36–8
Rosenberg, Alexander 145
Ruse, Michael 186, 194

Sandel, Michael 3, 8, 20–4, 39, 55–7, 61, 77
Scanlon, Thomas 149, 200–1
schooling, *see* education
sexual reproduction, *see* reproduction
Shea, Nicholas 88

siblings 132–3, 187
Singer, Peter 2, 125, 129–31
Sober, Elliott 99, 101, 140, 148, 154–6
social conditioning 129–31
social construction 130, 136, 179, 183
social engineering 114, 122
social learning 46–9, 88, 95–6, 137, 139, 144
social structural view (of justice) 149–50, 153–4, 156
sociobiology 128, 191
sociology 131, 138, 140, 142
Sommerhoff, Gerd 178–9, 182
species 39–43
  species as individuals 41–3, 48–9
  species design 2, 181, 184
  species' nature 150, 174
sperm 79–80, 82–6, 92, 111
spina bifida 18
Sterelny, Kim 18, 100, 107–8, 128, 144, 179
swampman 108
synthetic biology 3, 9, 60–71, 74–8

taboo 141–2
TaurOs Project 67
teleo-essences 52–3
teleosemantics 108
thalidomide 105
theological ethics 20
Thompson, Adrian 71–3
Thompson, Michael 51–4
Tooby, John 127, 133
Thornhill, Randy 125, 139–142, 188
transhumanism 17
treatment/enhancement distinction 175–7, 186–8, 199–203

udders 62; *see also* cattle
unconscious selection 67, 70
Universal Declaration on the Human Genome and Human Rights 6
utopianism 13, 28, 37–8

violence 130, 132, 134, 138–40

Waddington, Conrad Hal 98, 127
Wakefield, Jerome 188

Walton, John, Lord 5-6
wealth 37, 113, 141-2, 165, 168, 195
weaning 141
Weismann, August 6
welfare 97, 109-10, 119, 170, 176, 196, 199-201
well-being, *see* welfare

wheelchair users 119, 162, 178-81, 182-4, 198
Wilson, Margo 45, 125, 132, 134-9

yuck (as a reaction to new technology) 29-30, 38

zygotes 4-5, 7, 83, 112